Jo Ann Jamison

Late Woodland
Cultures
of the Middle Atlantic
Region

Late Woodland Cultures of the Middle Atlantic Region

Edited by Jay F. Custer

Newark: University of Delaware Press
London and Toronto: Associated University Presses

© 1986 by Associated University Presses, Inc.

Associated University Presses
440 Forsgate Drive
Cranbury, NJ 08512

Associated University Presses
25 Sicilian Avenue
London WC1A 2QH, England

Associated University Presses
2133 Royal Windsor Drive
Unit 1
Mississauga, Ontario
Canada L5J 1K5

The paper used in this publication meets the requirements of the American National Standard for Permanence of Paper for Printed Library Materials Z39.48-1984.

Library of Congress Cataloging-in-Publication Data
Main entry under title:

Late woodland cultures of the Middle Atlantic region.

   Bibliography: p.
   Includes index.
   1. Indians of North America—Middle Atlantic States—
Antiquities—Addresses, essays, lectures.   2. Woodland
Indians—Antiquities—Addresses, essays, lectures.
3. Middle Atlantic States—Antiquities—Addresses,
essays, lectures.   I. Custer, Jay F., 1955–
E78.M65L37   1986   974'.01   84-40807
ISBN 0-87413-285-1 (alk. paper)

Printed in the United States of America

Contents

List of Figures	6
List of Tables	7
Preface	9
Introduction JAY F. CUSTER	11
1 Difficulties in the Archaeological Identification of Chiefdoms as Seen in the Virginia Coastal Plain during the Late Woodland and Early Historic Periods E. RANDOLPH TURNER	19
2 Late Woodland Cultures of the Middle and Lower Delmarva Peninsula JAY F. CUSTER and DANIEL R. GRIFFITH	29
3 Late Woodland Cultures of the Middle and Lower Delaware River Valley and the Upper Delmarva Peninsula R. MICHAEL STEWART, CHRIS C. HUMMER, and JAY F. CUSTER	58
4 Cultural Diversity in the Lower Delaware River Valley, 1550–1750: An Ethnohistorical Perspective MARSHALL J. BECKER	90
5 Late Woodland Settlement Patterns in the Upper Delaware Valley HERBERT C. KRAFT	102
6 Late Woodland Cultures of the Lower and Middle Susquehanna Valley JAY F. CUSTER	116
7 Late Woodland Cultural Diversity in the Middle Atlantic: An Evolutionary Perspective JAY F. CUSTER	143
References	169
Notes on Contributors	207
Index	208

Figures

1. Regions covered in text — 14
2. Late Woodland settlement model no. 4 — 42
3. Late Woodland settlement model no. 5 — 43
4. Late Woodland settlement model no. 3 — 44
5. Selected sites of the Piedmont Uplands and Upper Delmarva Peninsula — 60
6. Abbott Farm location — 66
7. Late Woodland sites of the Abbott Farm locality — 68
8. Late Woodland settlement pattern no. 1 — 74
9. Late Woodland settlement pattern no. 2 — 75
10. Late Woodland settlement pattern no. 3 — 77
11. Selected Late Woodland sites of the Middle Delaware Valley — 81
12. Detail of Hugo Allard's map of 1673 — 110
13. Middle–Late Woodland ceramic distributions and interaction networks — 120
14. Shenks Ferry complex distribution — 122
15. Blue Rock/Stewart phase distributions and interaction — 126
16. Lancaster/Funk phase distribution and interaction — 127
17. Late Woodland cultural evolutionary trajectories in the Middle Atlantic region — 162

Tables

1. Slaughter Creek complex settlement models 33
2. Faunal remains: Slaughter Creek complex sites 45
3. Floral remains: Slaughter Creek complex sites 47
4. Mixture of Middle–Late Woodland ceramics from the Middle Delaware Valley 87
5. Lancaster lowland site data 129
6. Faunal food sources from Washington Boro midden 139
7. Plant food sources from Washington Boro midden 141

Preface

The papers that are included in this volume were solicited as part of a symposium entitled "Late Woodland Cultures of the Middle Atlantic," presented at the 1983 Middle Atlantic Archaeological Conference in Rehoboth Beach, Delaware. All of the papers except for one, that by Herbert C. Kraft, were presented at the conference. The overall goal of the symposium was to provide a current perspective and synthesis of the Late Woodland archaeological data from different portions of the Middle Atlantic. Participants were to go beyond the typical chronological concerns with projectile points and especially ceramics that characterized many earlier reviews of Late Woodland cultures. Specifically I asked participants to discuss settlement patterns, community organizations and inferred social organizations, subsistence systems—particularly the presence of horticulture—the use of storage, and the inferred level of socio-political organization. By emphasizing these topics I hoped that an understanding of the evolutionary processes that transformed the Early and Middle Woodland societies of the Middle Atlantic into the cultures observed at the time of European Contact could be developed. The Late Woodland period is an especially interesting time period for the study of these evolutionary processes in the Middle Atlantic because in some areas there is a clear-cut increase in social complexity, while in others there are good indications of either no social change or decreasing social complexity during this time.

The papers presented in this book are oriented toward the same goals. In many ways this book can be viewed as a follow-up to Roger Moeller's (1982) edited volume of environmental archaeology papers that included several papers from an Early and Middle Woodland session at the 1981 Middle Atlantic Archaeological Conference. Each paper presented here covers a specific geographical area and the last paper, an overview that briefly summarizes the Late Woodland archaeological record of areas not included in the other papers, offers some hypotheses and speculations about how and why cultures changed during Late Woodland times.

Introduction
JAY F. CUSTER

For more than twenty years it has been almost axiomatic that Late Woodland cultures—those cultures dating to the time period after A.D. 1000 and before the onset of European Contact circa A.D. 1600—were characterized by three major attributes throughout the Middle Atlantic region. These distinctive features were (1) settled village life; (2) use of agriculture; and (3) manufacture of ceramics with complex designs. In most older studies, these traits, especially the first two, functioned as true axioms; they were neither debatable nor demonstrable. For the most part researchers accepted them and studied the varieties of Late Woodland ceramics of the third attribute.

The development of these axioms and their acceptance were probably due to the fact that the best-known and earliest excavated Late Woodland sites in the Middle Atlantic region were sedentary agricultural villages such as the Susquehannock villages of the Lower Susquehanna Valley (Cadzow 1936), the Slaughter Creek complex sites of the lower Delmarva Peninsula (Davidson 1935), the large villages of the Lower Potomac River Valley (Schmitt 1952, 1965; Stephenson 1963), and the classic Iroquoian villages of New York State (Ritchie 1965; Ritchie and Funk 1973). In most of these cases it was clear that these sites were settled villages. They were very large and often contained abundant storage features and middens. Quite often defensive stockades were present as well, along with food remains of corn, beans, and squash, sometimes in substantial quantities.

Continued excavations between 1940 and 1960 showed no reason to question these axioms. Newer concerns, such as settlement pattern studies, were phrased within the traditional view of Late Woodland societies (for example see Kinsey 1975). Studies of the communities' internal settlement patterns became popular and much of the Late Woodland archaeological literature of this period contains maps of house patterns and village plans (for example see Ritchie and Funk 1973). However, some anomalous Late Woodland communities began to appear in regional settlement pattern studies in the Middle Atlantic region. For example, Jeff Graybill's study of Shenks Ferry settlement patterns in southeastern Pennsylvania (Graybill 1973; Kinsey and Graybill 1971) indicated that agricultural communities may have existed without settled village life and that the development of settled village life among Shenks Ferry populations was a result of conflicts with neighboring groups. These conflicts were viewed as late events in the Late Woodland chronology and the implication was that not all Late Woodland community members were sedentary agriculturalists living in large villages.

Settlement pattern studies on the Delmarva Peninsula by Thomas et al. (1975) also indicated that there were many community patterns in the coastal regions during Late Woodland times that were not based on sedentary villages. Even more important was the fact that there seemed to be good indications that there was a significant decrease in cultural complexity between Middle Woodland and Late Woodland times on the Delmarva Peninsula (Thomas 1977). It was no longer clear that all Woodland societies in the Middle Atlantic had followed the same evolutionary trajectory of increasing social complexity based on agriculture and increased sedentism, as the corollaries of the previously noted axioms would have had us believe. The variety of settlement patterns, community patterns, social organizations, and basic adaptations during Late Woodland times was much more complex than was originally thought. Clearly, there was need for something more than just ceramic studies.

These essays seek to provide an overview of the varieties of cultures that existed in the Middle Atlantic region between A.D. 1000 and the advent of European Contact, sometime in the seventeenth century. In one sense the authors try to demonstrate the variety of settlement and subsistence systems that contradict the first two axioms noted above. In some areas, there are clear-cut examples of sedentary agricultural villages while in others such lifeways never existed during Late Woodland times. In addition to merely documenting this variety, the papers also examine changes in Late Woodland cultures through time and try to provide some preliminary answers to the question of how and why these cultures changed. In other words, these papers provide a series of descriptions of the various multilinear pathways (Steward 1955) of sociocultural evolution during Late Woodland times in the Middle Atlantic region.

Many of the papers included in this volume do deal with some issues in ceramic typologies and chronologies. My earlier remark that something more than ceramic studies was needed should not be taken to mean that we no longer consider this aspect of Late Woodland material culture. However, studies of ceramic typologies and their chronological implications are not an end in themselves; they are

Introduction

merely tools that lead to more substantial issues. Nonetheless, as Stewart (1982d) has noted, typological studies need to be refined and reformulated in order to answer larger, more current research problems. In this volume, discussions of ceramic typologies are limited to chronological considerations that are germane to other issues and to stylistic analyses that can be linked to groups' interactions (Plog 1980).

Late Woodland Cultures is organized by drainages and geographical regions and begins with a paper by E. Randolph Turner on the relatively complex chiefdoms of the Virginia Coastal Plain. Turner's study is more problem-oriented than other papers in this volume and considers the archaeological correlates of known chiefdoms, such as the Powhatan Chiefdom. It also in effect provides an important cautionary note on the inference of social systems from archaeological data.

The Late Woodland cultures of the Lower Delmarva Peninsula are the subject of Jay F. Custer and Daniel R. Griffith's paper, which gives an example of reduction in social complexity between the Middle Woodland and Late Woodland periods. However, the societies of the Lower Delmarva Peninsula did develop a certain degree of social complexity and did use some small amount of agriculture by the onset of European Contact.

Moving north, R. Michael Stewart, Chris Hummer, and Jay Custer review the Late Woodland cultures of the Middle and Lower Delaware River Valley and the Upper Delmarva Peninsula. This area is characterized by a great deal of continuity between Middle Woodland societies and Late Woodland groups. Some of the simplest Late Woodland societies of the region are found in this area, particularly its southern reaches.

The paper written by Marshall Becker represents a slight departure from other contributions in this volume in that it uses a primarily ethnohistorical perspective to study the social complexity and diversity of groups living in the Lower Delaware River Valley during the initial stages of European Contact. By studying varied responses to European Contact among the Lenape, Becker shows that a band-level organization probably characterized the pre-Contact Lenape.

Continuing north up the Delaware River Valley, Herbert C. Kraft discusses the Late Woodland cultures of the Upper Delaware Valley. Kraft's paper considers the settlement patterns of the Upper Delaware and notes the agricultural subsistence base of the Late Woodland inhabitants of the area. It is interesting to note that large villages are not apparent in this region.

Custer next describes the Late Woodland cultures of the Lower and

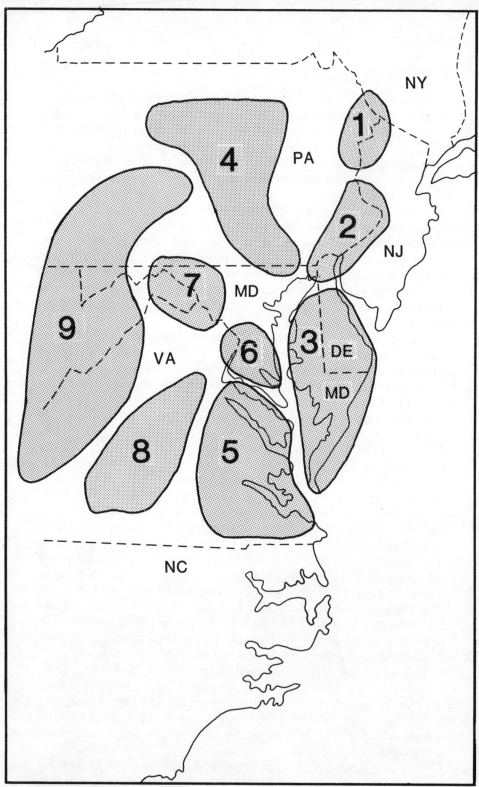

Fig. 1. Regions covered in text. 1: Upper Delaware Valley; 2: Middle and Lower Delaware Valley and Upper Delmarva Peninsula; 3: Middle and Lower Delmarva Peninsula; 4: Lower and Middle Susquehanna Valley; 5: Virginia Coastal Plain; 6: Potomac Coastal Plain; 7: Potomac Piedmont; 8: Virginia Piedmont; 9: Appalachian Highlands.

Introduction

Middle Susquehanna Valley and notes a variety of settlement and community patterns for this area, including large sedentary villages and small farmsteads. A series of interactions with more complex Iroquoian groups to the north had important effects on the development of social complexity in this region.

The final paper, an overview by Custer of the regions of the Middle Atlantic not covered by the other papers, also provides some hypotheses and speculations about how and why Late Woodland cultures changed in different ways in different areas within the Middle Atlantic region. The theories about tribal social organizations proposed by Elman Service, Marshall Sahlins, and Marvin Harris and Barbara Price are discussed in light of the Middle Atlantic Late Woodland data. A comprehensive reference list for all of the papers appears at the end of the volume.

Late Woodland
Cultures
of the Middle Atlantic
Region

1
Difficulties in the Archaeological Identification of Chiefdoms as Seen in the Virginia Coastal Plain during the Late Woodland and Early Historic Periods

E. RANDOLPH TURNER

One of the principal goals presently facing anthropological reasearch is the identification and explanation of observed variations in levels of sociocultural complexity. The study of this variation has resulted within recent years in the formulation of developmental sequences of societies ranging from basically egalitarian band and tribal societies through stratified state societies (cf. Service 1962; Fried 1967). Of particular concern in this paper is the archaeological identification of an intermediate stage known as the chiefdom society.

Identification of such a stage is especially important in archaeological discussions on sociocultural adaptation and evolution due to pronounced changes in economic, sociopolitical, and religious organization associated with the emergence of chiefdoms. As has been documented by Naroll (1956) and Carniero (1967), among numerous others, societal expansion in population size typically requires the rise of new means of sociocultural integration. In the emergence of chiefdoms, this development is manifested in the growth of ascribed positions of leadership with centers coordinating economic, sociopolitical, and religious organization. These leadership positions may be contrasted to the largely achieved positions of leadership characteristic of more egalitarian band and tribal societies. Given the greater and more marked variations in status found in chiefdoms, these societies ideally should be most easily detected archaeologically through variations in burial practices and spatial differentiations

---

I wish to thank James W. Hatch, William T. Sanders, Robert S. Santley, and David L. Webster for their valuable suggestions on an earlier version of this paper. More recent comments came from Jay F. Custer, William R. Gardner, W. Fred Kinsey III, Howard A. MacCord, and Stephen R. Potter, to whom I also am most grateful.

within and between settlements as expressed in community organization and settlement patterns.

Statements by Sanders and Price (1968), Diehl (1973), and Peebles and Kus (1977) show the significance of the two general categories just noted for the archaeological identification of chiefdoms in contrast to less complex societies. Speaking in general terms, Sanders and Price (1968; 115) note:

> Archaeologically, societies of this type are identified on the basis of site stratification. If, in a local area, there are numerous contemporary sites, some without civic architecture, and/or others with very small civic centers, and still others with markedly larger ones, then one can safely infer the presence of relatively large, stratified social systems involving a number of communities. When this evidence is combined with striking variations in the richness of tomb furniture, the archaeologist can assume the existence of a society in which social stratification—or at least ranking—is a major factor in integrating the social system.

Diehl presents a more specific list of six archaeological indicators for chiefdoms (1973; 19–20):

(1) A settlement pattern lacking large urban centers, though characterized by (a) ceremonial centers surrounded by villages, hamlets, or individual households, (b) a series of small "towns", or (c) a large single community which contains the majority of the population of the society.
(2) An economy which emphasizes intergroup exchanges of many items and intragroup trade in luxury goods but which lacks markets.
(3) Evidence of full- or part-time occupational specialists.
(4) Evidence of a ranked social structure divisible into two major groups, one dominant and the other subordinate.
(5) Striking intergroup differences in burial practices and mortuary offerings.
(6) Remains of public works such as temples, ceremonial precincts, and tombs.

Taking a somewhat different perspective, Peebles and Kus (1977; 431–33) discuss five major areas of variability that they suggest are characteristic of chiefdoms and whose presence or absence can be tested through archaeological data:

(1) There should be clear evidence of non-volitional, ascribed ranking of persons.

Difficulties in the Archaeological Identification of Chiefdoms

(2) There should be a hierarchy of settlement types and sizes, and the position of settlements in the hierarchy should reflect their position in the regulatory and ritual network.

(3) All other things being equal, settlements should be located in areas which assure a high degree of local subsistence sufficiency.

(4) There should be evidence of organized productive activities which transcend the basic household group.

(5) There should be a correlation between those elements of the cultural system's environment which are of a frequency, amplitude, and duration to be dealt with but which are least predictable and evidence of society-wide organizational activity to buffer or otherwise deal with these perturbations.

Unfortunately, identificatory characteristics such as noted in the above examples by Sanders and Price, Diehl, and Peebles and Kus frequently are difficult to detect, which leads to potential errors in the interpretation of the archaeological record. In this paper I describe one such case from the Virginia Coastal Plain from which may be drawn significant implications regarding the archaeological recognition of both low-level and more complex chiefdoms.

The Powhatan Chiefdom

For the early 1600s, the initial years of English settlement in the Virginia Coastal Plain, extensive ethnohistorical data confirm the presence of a relatively complex rank society in this region known as the Powhatan Chiefdom. By this time, the Powhatan were at their height in terms of internal centralization and external expansion. The following summary on the Powhatan as a chiefdom is based principally on historical documents edited by Arber (1910), Barbour (1969), and Wright and Freund (1953), and anthropological interpretations of such data by Binford (1964) and Turner (1976, 1982a).

The origin of the Powhatan Chiefdom can be traced back to sometime before the mid- to late-1500s, at which time Powhatan inherited six to nine districts along the James and York rivers. As seen in historical documents edited by Quinn (1955), during the 1580s the English described a series of small low-level independent chiefdoms along the North Carolina coast that attempted to expand their range of jurisdiction through alliances and warfare. A similar situation was encountered by Powhatan when he attained control of his inherited territories. By 1607 the chiefdom had expanded principally through warfare or threat of warfare to include approximately thirty-one districts, or subchiefdoms, encompassing roughly 16,500 square kilometers and 13,000 persons. At this time, the year the English settlement of Jamestown was founded, English documents describe a remarkably complex chiefdom, especially notable if one recognizes the rapid expansion of the chiefdom just prior to English contact.

As a point of comparison, many of the Powhatan Chiefdom's characteristic features are similar to ones in some of the more complex African and Polynesian chiefdoms discussed in studies by Taylor (1975) and Sahlins (1958). I give brief examples of such features below. Note, however, that by making such a comparison and referring to Powhatan society as representing a relatively complex chiefdom, I am not implying that the Powhatan Chiefdom is equivalent to the most complex chiefdoms identified archaeologically or ethnographically. Recently Carniero (1981; 47) has proposed a three-level typology of chiefdoms:

> A *minimal* chiefdom is one that meets the minimal requirements of a chiefdom but does not go far beyond them. A *typical* chiefdom is one that is clearly a chiefdom, with elaborations in many aspects of its political and social structure, but still well below the level of a state. A *maximal* chiefdom is one that has become large and complex enough to approach the threshold of the state.

The Powhatan Chiefdom clearly fits Carniero's definition of a typical chiefdom, although some features align it more closely with maximal chiefdoms than with minimal chiefdoms.

Three major status levels—paramount chief, district chief, and nonchief—characterize the Powhatan society. The nonchief level further divides into priests and shamans, counselors and distinguished warriors, and finally commoners. A complex redistributive hierarchy was in operation, and chiefs were able to confiscate specific goods of others. Such accumulated goods principally supported visiting dignitaries and various communal feasts and religious activities; aided groups within the chiefdom during times of famine; and obtained allies and rewarded courageous warriors in warfare. Formally defined tribute labor planted and harvested at least the paramount chief's agricultural fields. Insignia of rank were observed through a significant range in clothes and ornaments, and chiefs had their own attendants, body guards, and possibly orators. Elaborate obeisance postures and other forms of respect were also present. The English accounts noted, in addition, the presence of what Sahlins in particular calls "arbitrary despotism" and control of socio-regulatory processes by high chiefs with marked status differences and having the ability to inflict secular punishment on wrong-doers. Data on unique rites held on a spectacular scale for all life crises of high chiefs are somewhat limited, although the ceremony of Huskanawing, probably a puberty ceremony for restricted individuals, involved a complex series of rites, and chiefs were treated in a particularly respectful manner upon their deaths. At this time their corpses underwent an elaborate process of preservation and were placed in special temples; along with various artifacts denoting rank.

Further indicating a complex system of rank was the restricted access placed on temples and storehouses under the control of a chief, since usually only chiefs and priests could visit these structures. In addition, chiefs' houses were larger than those used by persons of lesser rank. Finally, within a complex chiefdom one would expect a well-defined hierarchial arrangement of settlements. This pattern is clearly indicated on Smith's (Arber 1910, 384) 1612 map of the region. Most districts are represented not only by hamlets and minor villages but also by a district "capital," which Smith labeled "kings howses." The center or core area of the chiefdom in terms of population distribution and many sociopolitical and religious activities was located at the confluence of the Pamunkey and Mattaponi rivers, where four such "kings howses" are situated in extremely close proximity to each other.

At the same time, Powhatan society also revealed other conservative features. For example, no evidence exists for extensive direct supervision of household production, and chiefs' decrees on the utilization of land, riverine, or marine resources were uncommon. Chiefs consulted with priests and counselors on important decisions such as warfare issues. Furthermore, the position of counselor appears to have been an achieved status position, although the position of chief, and perhaps priest as well, was ascribed. Precise matrilineal inheritance rules determined who became chief. Finally, there is a noted absence of full-time craft specialization in Powhatan society and only minimal evidence for even part-time craft specialization.

Regardless of these conservative features, it still remains clear that Powhatan society in 1607 represents a relatively complex chiefdom. This fact is well-documented in previously summarized descriptions of status levels, redistributive systems, insignia of rank, forms of sociopolitical control, burial practices, specialized structures, and a hierarchial arrangement of settlements.

One would expect that the Powhatan Chiefdom could be easily identified archaeologically as a chiefdom through such characteristics as:

(1) The presence of specialized structures, namely chiefs' houses, storehouses, and temples;
(2) Differential distribution of restricted rank-denoting artifacts (e.g., copper artifacts, pearls, and certain shell artifacts) within settlements, between settlements, and within burials; and

(3) A settlement hierarchy similar to the one described by Smith for the Powhatan.

However, just the opposite is the case. If all historic records on the Powhatan are disregarded, an archaeological reconstruction of Late Woodland Period aboriginal life within the region results in an overview quite similar to proposed lifeways within such present-day states as Maryland, Delaware, and New Jersey. At best, only limited, nonconclusive data exist concerning the presence of rank societies in the Virginia Coastal Plain.

Published material on archaeological research in coastal Virginia dates back to as early as the late 1800s. At present, published reports contain data for well over one-hundred prehistoric sites in the region. In addition to these reports on individual site excavations and surface collections, further data are available in summaries of several extensive archaeological surveys conducted in portions of the Virginia Coastal Plain occupied by the Powhatan. Although the data are limited, the region is still no less well represented archaeologically (including specifically the Late Woodland and Early Historic periods) than many, if not most, other areas of eastern North America.

For instance, numerous examples exist of limited and more intensive test excavations on archaeological sites with Late Woodland/Early Historic components on all major river drainages in the Virginia Coastal Plain. Representative reports include those by Buchanan (1966, 1969), McCary (1967), MacCord (1967), and Reinhart (1978) for the James River; McCary (1953) and MacCord and Owens (1965) for the Chickahominy River; MacCord (1964a, 1975) for the Appomattox River; McCary (1958c) for the York River; Winfree (1967, 1969) for the Mattaponi River; Owens (1969) for the Pamunkey River; MacCord (1974) for the East River; MacCord (1965, 1969) for the Rappahannock River; and Buchanan (1976), Potter (1982), and Waselkov (1982) for the Potomac River. Supplementing the above examples are intensive excavations at such likely prehistoric villages are the Hatch site (Gregory 1980) and Flowerdew Hundred (Charles T. Hodges, personal communication) on the James River, the Great Neck site on Broad Bay in Virginia Beach (unpublished data on file at the Virginia Research Center for Archaeology), and the Patawomeke site on the Potomac River (Schmitt 1965). Examples of major surveys range from early studies such as that by Bushnell (1937) to more recent research by Gardner (1981), McCary and Barka (1977), Potter (1982), Turner (1976), and Wittkovski (1982a, 1982b), among others.

A review of this material indicates that no specialized structures have been identified archaeologically that could be considered as chiefs' houses, storehouses, or temples. Similarly, available survey data show no clear hierarchical arrangement of settlements during Late Woodland times. Variations in settlement size that do exist are best explained through seasonal subsistence exploitation practices not necessarily characteristically of chiefdoms alone.

The presence of valued nonutilitarian items possibly denoting ascribed rank occurs only in five cases, all of them involving burials. Typical burials, either in ossuaries (Norman F. Barka, personal communication; Barka and McCary 1969; Bushnell 1920, 28–29; Bushnell 1940, 142–43; Mann 1981; T. D. Stewart 1940b) or in individual graves (Gregory 1980 and personal communication; McCary 1967, 69–70; MacCord 1964b, 81; MacCord 1967, 76–77; MacCord 1975, 16; MacCord and Owens 1965, 82), are associated with few or no artifacts. At two sites (Patawomeke and an adjacent site) on the Potomac River and one site (Mount Airy) on the Rappahannock River large numbers of artifacts, including shell products and copper as well as occasional early historic artifacts, are associated with burials (McCary 1950, 1958a, 1958b; Potter 1982, 101–2, 105–6; Reynolds 1883, 93; Schmitt 1965, 20; T. D. Stewart 1939, 87; T. D. Stewart 1940a, 82; T. D. Stewart 1941, 70). The lack of specific data on accompaniments per burial prohibits further statements except that the overall number and types of artifacts found indicate possible ranked statuses. Burials at these sites appear to date from the Late Woodland to Early Historic period. A similar situation occurs at the Great Neck site adjacent to Broad Bay in Virginia Beach, although here burials date principally to the Late Woodland period, possibly extending back as early as the Early Woodland period (Painter 1980; unpublished data on file at the Virginia Research Center for Archaeology). Finally, a recently reported Late Woodland period burial at the Hatch site on the upper James River (P. W. Peebles 1983, 2) may represent another example of a ranked individual based on the unusually large number of shell artifacts present. Of particular importance is the fact that all of the above sites are outside of the core area of the Powhatan Chiefdom and not where one would have predicted their sole occurrence. Further, by themselves (given available data) they simply are not sufficient evidence to confirm clearly the presence of ranked societies in the region.

Ethnohistorical data are quite clear that the evolution of the Powhatan Chiefdom from a series of low-level chiefdoms and tribal societies into a relatively complex rank society was not the result of English contact. That the Powhatan Chiefdom predates the Historic period is evident in English descriptions dating to 1607 of aspects of sociopolitical and religious organization clearly characteristic of a well-developed chiefdom. Similarly, other likely contacts with Europeans during the preceding sixteenth century were minimal, with changes in Powhatan sociocultural complexity principally resulting from factors indigenous to the local region (cf. Binford 1964; Turner 1976; Turner 1982a; Potter 1982). At the same time, this transition took place over a very short period of time—that is, during Powhatan's lifetime—making archaeological identification extremely difficult. Similar problems hinder identifying the low-level chiefdoms out of which the Powhatan Chiefdom evolved since such societies leave few archaeological traces that would distinguish them from more egalitarian tribal societies.

Thus, without extensive regional surveys and accompanying excavations, which are typically beyond the financial capabilities of most research projects, the presence of prehistoric chiefdoms presents specific difficulties in identification since diagnostic features may be limited in number and often hard to locate. Archaeological data from the Virginia Coastal Plain indicate that this applies specifically to two cases: (1) low-level chiefdoms, where forms of rank distinction are not well established, and (2) more complex chiefdoms of a short duration. Similar problems also characterize other portions of Virginia (Turner 1981; Gardner 1982).

Such identification problems are more pronounced when examples of monumental architecture are absent. This is clearly the case for the Virginia Coastal Plain. No mounds, whether they served for burials or bases for temples or residential structures for highly ranked individuals, have been clearly documented archaeologically. While there are two reported cases of possible mounds near the confluence of the Pamunkey and Mattaponi rivers in the core area of Powhatan Chiefdom (Richardson 1884, 828; Speck 1928, 302), they have never been verified archaeologically. Similarly, there are no other forms of conspicuous structures of a residential socioeconomic (e.g., storage) and/or religious function with above-ground traces. In contrast, where such specialized structures can be more readily recognized, the archaeological identification of chiefdoms is less difficult, as examples show from other portions of the eastern United States as well as Mesoamerica and South America (Sanders and Marino 1970). For examples related specifically to the eastern United States, see Brown 1971; Gibson 1974; Hatch 1976; Larson 1971; C. S. Peebles 1971; and Seeman 1979, among others.)

Discussion and Summary

Recently there has been a growing interest in the archaeological recognition of varying levels of sociocultural complexity. This paper has been concerned with one such level, the chiefdom society, and the archaeological indicators used in identifying it. Based on archaeological and ethnohistorical data on the Powhatan and their predecessors in the Virginia Coastal Plain, I have noted two problems, one involving low-level chiefdoms and the other more complex chiefdoms. Similar problems undoubtedly plague numerous other prehistoric cultures previously identified as representative of egalitarian societies. Several examples can be given for eastern North America.

Winters presents data on artifact distribution in Late Archaic period Indian Knoll burials in Kentucky that indicate possible incipient low-level chiefdoms, but then he notes, ". . . in other respects the archaeological data suggest that the population was socially homogeneous and without great differentials in the basic way of life of the individuals composing the population" (1968, 209). If such Late Archaic societies indeed represent low-level chiefdoms (see also Rothschild 1979), this recognition would be of considerable importance in archaeological attempts to delineate the evolutionary processes operative in the development of complex rank societies in the general region by Hopewellian times.

Similarly, ethnohistorical data on early contact societies in southern New England, such as the Narrangansett, Wampanoag, and Massachusett, show evidence of low-level chiefdoms in this region, as seen by scattered references on ascribed positions of leadership, redistribution systems, variations in dress, house size, burial practices dependent upon rank, prescribed marriage rules for chiefs, and the presence of specialized temple structures characterized by limited access. Many of these sources are listed in Rainey's 1956 review of the ethnohistory of the area. However, prehistoric archaeological evidence for the presence of rank societies in southern New England is noticeably absent. Their identification as low-level chiefdoms is particularly important since aboriginal societies in northeastern North America typically have been classified in the past as more egalitarian bands or tribes. Moreover, even if a result of European Contact, the recognition of low-level chiefdoms in portions of this region poses new research problems regarding changing sociocultural adaptations over time in the Northeast.

Finally, in a series of recent papers, Custer (1982b, 1983b; Custer and Stewart 1983) has hypothesized the presence of incipient rank societies, though not necessarily chiefdoms, in the Delaware Coastal Plain by the Middle Woodland period that did not continue into the Late Woodland period. While I do not dispute the possibility of such being the case, the specific data presented by Custer have been limited to date. A similar situation characterizes the arguments raised by R. M. Stewart (1982c) for possible incipient rank societies during the Middle Woodland period in the New Jersey Coastal Plain. Note too that Stewart and Custer both identify these Middle Woodland societies as "ranked societies" rather than "chiefdoms." Although Fried's "ranked societies" (1967) typically are considered equivalent to Service's "chiefdoms" (1962), Taylor (1975, 16; see also Carniero 1981, 44) has argued that differences do exist, with some low-level rank societies not strictly meeting Service's definitional characteristics of chiefdoms. While outside the scope of this paper, the archaeological identification of such societies, whether representative of chiefdoms or not, is important in documenting the continuum of societies ranging from basically egalitarian bands and tribes through more complex chiefdoms. If either Custer or Stewart is correct, then apparently different evolutionary trajectories were occurring in the Delaware and New Jersey Coastal Plain in comparison to the Virginia Coastal Plain (compare references cited above with evolutionary models presented by Binford [1964] and Turner [1976]). While not unexpected (cf. K. V. Flannery, 1972; Sanders and Webster 1978), the delineation of such trajectories is critical in order to more precisely understand the adaptive processes underlying the development of rank societies.

As the above examples show, the proper identification archaeologically of rank societies such as chiefdoms is essential to avoid otherwise unnecessary misinterpretations of the archaeological record on changing sociocultural adaptations occurring over time. Unfortunately, in many cases this is hard to do. This situation is illustrated in the inability to identify clearly Powhatan society as representing a relatively complex chiefdom, much less merely a rank society, if one is using only existing archaeological data. By raising this problem, I am in no way recommending that we lower the standards for identifying chiefdoms archaeologically, which is obviously counterproductive. However, it is essential to project viable hypotheses, such as was recently done by Custer and Stewart, and then explicitly recognize that verification of such hypotheses will likely require integrated large-scale archaeological data recovery programs in the form of both extensive surveys and excavations not typically characteristic of the region today.

2
Late Woodland Cultures of the Middle and Lower Delmarva Peninsula

JAY F. CUSTER and DANIEL R. GRIFFITH

The Late Woodland cultures of the Middle and Lower Delmarva Peninsula exhibit a variety of cultural adaptations across both time and space. This paper describes this variation and explains the sociocultural evolutionary processes that cause it. Specifically, we consider the societies living in the southern two-thirds of the Delmarva Peninsula between A.D. 1000 and A.D. 1600. In keeping with the trends addressed in other papers in this volume, we stress archaeological data pertaining to settlement-subsistence systems, social organizations, and sociocultural complexity.

The Late Woodland chronology for the southern two-thirds of the Delmarva Peninsula is based on two main sources of data: seriation of Townsend ceramics and radiocarbon dates. In general, the Late Woodland cultures of this area are subsumed within the Slaughter Creek complex as originally defined for southern Delaware by Thomas (1973, 1977) and expanded upon by Custer (1983a; chap. 5). Recent overviews of the late prehistoric archaeology of the Eastern Shore of Maryland (Hughes 1980; Davidson 1982a; Custer 1983c) and Virginia's Eastern Shore (Wittkovski 1982a) also have shown that the Late Woodland societies of these areas can be included within the Slaughter Creek complex. The defining characteristics of this complex include the presence of Townsend ceramics, triangular projectile points, and large semipermanent or permanent base camps with a number of associated storage, refuse, and processing features (Thomas 1973, 1977).

Townsend ceramics, the major defining criterion of the complex on the Delmarva Peninsula, were originally defined by Blaker (1963) as shell-tempered, fabric-impressed, conoidal vessels with varying kinds of incised and corded decorative motifs. Griffith's (1977, 1982; Griffith and Custer n.d.) detailed analysis of design motifs within the Townsend series of ceramics provides a basis for a seriation of designs that can roughly distinguish between early Late Woodland sites (pre-A.D. 1350) and later Late Woodland sites (post-A.D. 1350). Simply stated, Townsend designs are more complex during earlier periods than during later ones. Early designs exhibit more design fields (Shepard 1954) and a wider range of broadline-incised geometric design motifs (Griffith 1982, 62, types RI4–RI8). Later designs are primarily corded with motifs composed of simple horizontal bands (Griffith 1982, 62, types RI1–RI2). Similar trends in design motif complexity over time can be seen in related ceramic wares, such as Minguannan and Overpeck, in the Lower and Middle Delaware River Valley (Griffith and Custer n.d., table 3).

Radiocarbon dates also support the Griffith seriation (Custer 1983a, appendix 2). Early complex ceramic designs are associated with the following radiocarbon dates: A.D. 975 (975 ± 60 B.P., SI-4946)—Slaughter Creek site (7S-C-30c); A.D. 1085 (865 ± 75 B.P., UGa-923)—Mispillion site (7S-A-1); A.D. 1100 (850 ± 55 B.P., UGa-1440)—Bay Vista site (7S-G-21); A.D. 1270 (680 ± 50 B.P., SI-4944)—Slaughter Creek site (7S-C-30b); and A.D. 1285 (665 ± 75 B.P., UGa-925)—Warrington site (7S-G-14). Later corded horizontal designs are associated with the following radiocarbon dates: A.D. 1345 (605 ± 50 B.P., SI-4943)—Slaughter Creek site (7S-C-30a); and A.D. 1370 (580 ± 60 B.P., UGa-924)—Poplar Thicket site. Thus there are two time periods within the Late Woodland Slaughter Creek complex of the Middle and Lower Delmarva Peninsula.

Two archaeological complexes were the immediate precursors of the Slaughter Creek complex—the Webb complex in the central and northern portions of the peninsula and the Late Carey complex in the lower peninsula. Each of these complexes dates from A.D. 500 to A.D. 1000 and shares certain basic similarities in adaptations. However, there are some important differences between the two. Settlement-subsistence patterns of the Webb complex and the Late Carey complex are quite similar. Both are based upon hunting and intensive gathering subsistence bases (Griffith 1974; Custer 1982b, 33) and exhibit two basic functional site types: habitation and procurement sites. Unlike the preceding Middle Woodland and Early Woodland complexes of the Delmarva area, habitation sites of the Webb and Late Carey complexes are all comparatively small and can be termed microband base camps (Custer 1983a). Only one macroband base camp, the Hell Island site (Thomas 1966a), is clearly evident from the archaeological record for either complex. This settlement pattern comes from a fissioning of large communities that occurred prior to Webb complex and Late Carey complex times (pre-A.D. 500) due to localized population pressures and amelioration of the environmental stresses that accompanied the midpostglacial xerothermic (Custer 1982b, 30).

The most significant difference between the two complexes is the association of one major mortuary center and three possible supralocal exchange centers with the Webb complex. The Island Field site (7K-F-17), which provided the original descripton of the Webb complex (Thomas and Warren 1970a), is a major cemetery site located in Kent County, Delaware. More than one hundred burials have been excavated to date and a wide variety of burial treatments and grave good assemblages were present. Thomas (1973, 1974) suggests that the variation in grave goods associations may indicate ranked kin groups, and Custer has also hypothesized the presence of a ranked society (1982b, 35–36). Some of the grave goods, including platform pipes and large pentagonal bifaces from the Island Field site, were manufactured from nonlocal materials (Thomas and Warren 1970a, 18).

Artifacts similar to those from the Island Field burials made from similar nonlocal materials are known from other sites on the Delmarva Peninsula (Thomas 1966a; Custer and Doms n.d.; Jackson 1954). These sites are exceptionally large and rich in artifacts compared to other local sites and may be the focal points of local social groups. Furthermore, analysis of quantified regional predictive settlement models for the regions surrounding the Hell Island site (Wells, Custer, and Klemas 1981) and the Island Field site (Eveleigh, Custer, and Klemas 1983) shows that neither site is located in the most optimal environmental zones, as are most large sites in the region. Rather, the location of sites like Island Field and Hell Island seem to be central to those of other contemporary sites in the region. This pattern indicates a big-man, supralocal social organization for some areas during Webb complex times. The absence of any single contemporary living site that can be directly associated with the Island Field cemetery (Custer 1982b, 32–33) also underscores this hypothesis. No cemeteries or comparable nonlocal exchange items are noted for any Late Carey complex sites.

In sum, during late Middle Woodland times, Webb complex societies of the Delmarva Peninsula were involved in exchange networks and other forms of social interaction with groups to the north and northwest. Some ranked social organizations were present to organize supralocal community maintenance of special mortuary and exchange center sites. In the southern peninsula region, similar complex social organizations and supralocal interaction patterns were not present among Late Carey complex groups. Nonetheless, both complexes shared a similar hunting and intensive plant food gathering subsistence base.

Settlement Patterns

When considering the Late Woodland societies of the Delmarva Peninsula, a researcher necessarily focuses on the archaeological record of southern Delaware because most of the excavated site data come from this area. Also, within the southern Delaware area there is sufficiently detailed information on Townsend ceramic designs from excavated sites to allow the application of the Griffith seriation. However, other sites can be considered in discussing Late Woodland settlement patterns on the Delmarva Peninsula. In this section we consider general Late Woodland settlement patterns.

The general adaptations of the Slaughter Creek complex have been studied intensively by Thomas et al. (1975), who analyzed the potential food sources found in the different environmental zones of southern Delaware and were able to develop a series of models of archaeological site distributions for the groups of people that would be exploiting these food resources. They noted two basic site types, including seasonal camps and base camps (Thomas et al. 1975, 62). Base camps included numerous storage/processing features, a wide variety of tool types, and possible structures, and would correspond to macroband base camps. Seasonal camps are smaller, contain fewer features, and would correspond to microband base camps (Gardner 1982). Thomas made no projections of specific procurement site locations. From this analysis, the researchers generated five basic settlement pattern models, each of which projected varied combinations of microband and macroband base camps in different environments during different seasons. Table 1 summarizes each model. Each model assumes a different degree of residential stability, ranging from groups of transient microband base camps to single sedentary macroband base camps or villages. After the models were developed, Thomas noted the expected artifact distributions for each model. In this manner the expectations of prehistoric activities for each model could be compared to the actual distributions of artifacts recovered in the archaeological record. After testing, the model whose expectations were most similar to the observed artifact distributions could be considered the most accurate picture of the prehistoric adaptations.

Models 3, 4, and 5 (see table 1) were the most accurate when the various settlement models were compared to two known sites in a preliminary test. These models, which have the highest degree of residential stability, suggest that in moving from the Middle Woodland to the Late Woodland period, lifestyles become more sedentary.

Table 1
Slaughter Creek Complex Settlement Models

Model	Winter	Spring	Summer	Fall
1	microband base camp; interior	microband base camp; mid-drainage	microband base camp; coastal	microband base camp; mid-drainage
2	macroband base camp; interior	macroband base camp; mid-drainage	macroband base camp; coastal	macroband base camp; interior
3	macroband base camp; interior	macroband base camp; coastal		macroband base camp; interior
4	macroband base camp; mid-drainage		microband base camp; coastal	macroband base camp; mid-drainage
5	macroband base camp; mid-drainage			

SOURCE: Thomas et al. 1975, 60–65.

Unfortunately, the available data are not really sufficient to discriminate among the three most accurate and most sedentary models. Nonetheless, we use the various site types and locations noted in these models to organize the discussion of the Slaughter Creek complex sites.

Macroband Base Camps

Macroband base camps and possible villages of the Slaughter Creek complex are found at a number of locations throughout southern Delaware and the rest of the lower peninsula. A series of macroband base camps and small village locations along Slaughter Creek provided the original data for the description of the Slaughter Creek complex (Davidson 1935a, 1935b, 1936; Purnell 1958; Zukerman 1979a, 1979b; Weslager 1968, 58–61). The sites from the Slaughter Creek location contained a wide range of tool types (especially plant processing tools), numerous varieties of faunal remains (Davidson 1936), and a possible semisubterranean pit house feature (Artusy and Griffith 1975). Smaller site areas in the vicinity include individual clusters of shell-filled pits and possible pithouses (Zukerman 1979a, 1979b) and may represent hamlets related to the main site (Custer

1983a). These varied sites may, or may not, be contemporaneous: radiocarbon dates from the sites range between A.D. 975 and A.D. 1345 (Custer 1983a, appendix 2).

An additional feature type found in the original Slaughter Creek excavations (Davidson 1935a, 1935b) was a large pit that included a number of human skeletons. Some of the skeletons were single fully articulated burials while others were in groups and appear to be secondary burials with many disarticulated individuals. Davidson (1935b) described this burial feature as an "ossuary." Similar burial features were also discovered at Late Woodland sites elsewhere in southern Delaware, including a series of burials near Rehoboth Beach excavated by Wigglesworth (1933), Thompson's Island burials excavated and reported by Weslager (1942) and T. D. Stewart (1945), and a series of sites along the Choptank on the Eastern Shore of Maryland (Weslager 1942). Thomas (1973) reviewed many of these burial features and notes that many seem to show some secondary treatment of the skeleton to produce the disarticulated remains. In a more recent review, Ubelaker (1974) writes that these sites on the Delmarva Peninsula are probably not true ossuaries like the large accumulations of individuals seen at sites in the lower Potomac River Valley and recorded among various Algonkian groups of the eastern seaboard (Feest 1973). Nevertheless, some special reburial of people seems indicated by the Slaughter Creek data.

The Townsend site (7S-G-2) near Lewes, Delaware, is another large Slaughter Creek complex macroband base camp or village excavated for several reasons by the then newly formed Sussex Society for Archaeology and History (Omwake and Stewart 1963). More than ninety shell-filled pits were excavated and the archaeologists found a disturbed grave feature including approximately nineteen individuals in both complete and disarticulated burials (Omwake and Stewart 1963, 4–5). Also recovered were shells from the features including oysters, clams, conch, and mussel (Omwake and Stewart 1963, 6), as well as numerous bone specimens, including tools.

The dating of the Townsend site is enigmatic. The ceramics described by Blaker (1963) include both Townsend-corded and Townsend-incised designs, which indicates the presence of the entire 600-year time range of the Slaughter Creek complex (Griffith 1977). Omwake and Stewart note some artifacts from early colonial times from the site (1963). Roulette-decorated pipes found in the feature fill are often found at sites that date to post-European Contact (after 1620). Plotting of the distribution of the Townsend designs across the site did show that there are two distinct clusters of features. One group includes the earlier Townsend designs and another set comprises the later corded designs and the roulette-decorated pipes. Thus there seem to be two occupations of the Townsend site: one prior to A.D. 1300 and another that may be as late as the A.D. 1550 date proposed by Witthoft (1963).

The Mispillion site (7S-A-1) represents another large Slaughter Creek complex macroband base camp or possible village and is very similar to the Townsend and Slaughter Creek sites. The Mispillion site has a long history of excavation (Hutchinson 1955a, 1955b; Hutchinson et al. 1957; Omwake 1954a; Flegel 1959; Thomas and Warren 1970b; Tirpak 1978). Many shell-filled pits have been noted and the wide variety of tool types and the large number of ceramics seem to resemble the other large Slaughter Creek complex base camp sites. Artusy and Griffith (1975) note that a semisubterranean house feature was also present at Mispillion. Lopez's 1961 report on the ceramics from the site shows a predominance of incised designs, and he dates the site to the earliest portions of the Slaughter Creek complex. Seriations of Townsend ceramics also place this site early in the sequence of the Slaughter Creek complex sites (Griffith 1977, 129–30, and a single radiocarbon date of A.D. 1085 supports this contention.

One of the more recently excavated Slaughter Creek complex macroband base camps is the Hughes-Willis site (7K-D-21) in central Kent County, Delaware. The Hughes-Willis site is the northernmost and smallest site of the Slaughter Creek complex and was excavated by the Delaware Bureau of Archaeology and Historic Preservation in conjunction with the Kent County Archaeological Society. There were some Middle Woodland Carey complex components at the site; however, the major part of the occupation is from Slaughter Creek complex times. The Hughes-Willis site is one of the sites used to test the various settlement models for Late Woodland groups (Thomas et al. 1975, 70–78). Detailed analysis of the tool categories revealed a wide variety of tool types. Ten features were excavated, representing about 20 percent of the total site. Faunal and floral remains from the site (Thomas et al., 1975, 76) indicate a fall through midwinter occupation. Extensive hickory nut processing also seems to have taken place at the site itself (Thomas et al. 1975, 77).

Other macroband base camps and possible villages have been discovered elsewhere on the Delmarva Peninsula; however, most of these sites are poorly known archaeologically. A few of them are along the lagoons and bays behind the Atlantic Coast barrier island complexes of the Delmarva Peninsula; however, they are not common. These sites include the Russell site (7S-D-7) reported in Delaware by Marine (1957); the Buckingham site (18W011), which is listed on the National

Register of Historic Places (Maryland Historical Trust 1974a—hereafter referred to as MHT); and the Tizzard Island site (18WO85), which is noted in the site files of the Division of Archaeology, Maryland Geological Survey. (Hereafter this site data will be referred to as MGS site files.)

In the southern portion of the peninsula macroband base camps of the Slaughter Creek complex are more common in the floodplains of the major drainages of the Chesapeake Bay on Maryland's lower Eastern Shore. Most of these sites are located close to the oligohaline (the freshwater/saltwater interface) in areas with extensive marshes. Sites included are Moore (18DO13), reported by Callaway, Hutchinson, and Marine (1960); Egypt Road (18DO36), noted by Custer (1983c; table 16); Nassawango (18WO23), which is noted as a major Late Woodland settlement focus by Hughes (1980; 194–217); and a series of four large sites on the Chicamacomico River, a tributary of the Nanticoke. Included among the Chicamacomico sites is the Brinsfield I site (18DO24), which is listed on the National Register of Historic Places (MHT 1974b), and the Lankford site (18DO43). Numerous shell-filled pits were excavated by Flegel (1975a, 1975b, 1976), and Griffith provides a summary of the site's ceramics (1977).

In addition to the floodplains of major drainages, bluffs overlooking embayed sections of drainages with drowned lower terraces and extensive marshes are important locations for macroband base camps along the Chesapeake Bay drainages. The Wessel site (18CA51) (Bastian 1974; Custer 1983c; table 16; MGS site files), excavated during an annual field school of the Archeaological Society of Maryland, is a good example of base camps in this environmental zone. Other macroband base camps on bluffs overlooking marshes include Long Point (18SO21); Spring Branch (18CA12); Tuckahoe Spring (18CA31); Tuckahoe (18CA39); Fowling Creek (18CA46); 18KE228; and 18CA51 (Custer 1983c, table 16; MGS site files). The Occohanoc site (44NH3) on Virginia's Eastern Shore is also in a similar environmental setting (Virginia Research Center for Archaeology site files; hereafter referred to as VRCA site files).

Microband Base Camps

Most of the Slaughter Creek complex sites now known fall into the microband base camp category. In Delaware, a few of these sites were the focus for some of the state's earliest archaeology (Weslager 1968, 10–30) and most are located along marshes, lagoons, and bays of the Delaware Bay and Atlantic Coast. A series of small-shell middens in the vicinity of Rehoboth Beach and Lewes were discovered in the nineteenth century by Leidy (1865) and Jordan (1880, 1895, 1906). Very little is known about these sites although their ceramics fall within the Townsend description. The small size of some of the sites attests to the fact that they are quite different from the larger Slaughter Creek complex sites described earlier. Over the years, similar sites have been discovered in approximately the same area, including the Rehoboth Midden (7S-G-3) reported by Marine et al. (1965) and the Lighthouse site complex (Weslager 1968, 26; Delaware Division of Historical and Cultural Affairs 1976). Especially interesting are the Lighthouse site and an additional shell midden (7S-D-22,9) that represent the latest of a series of Early/Late Woodland and microband base camps, or especially large and intensively utilized procurement sites, that are associated with the changing estaurine environment behind the Cape Henlopen spit (Kraft and John 1978). The continuity of site locations through the Woodland period in these brackish water environmental zones underscores the similar nature of at least some aspects of the subsistence activities between Early/Middle Woodland and Late Woodland times.

Another close correspondence of locations and apparent function between Early/Middle Woodland and Late Woodland microband base camps is at the Millman site complex in Kent County, Delaware (7K-E-4,23) (Thomas 1966b; Delaware Division of Historical and Cultural Affairs 1980). Webb and Slaughter Creek components overlap and indicate continuity of Middle/Late Woodland adaptations in this particular area in the northern portion of the Slaughter Creek complex. Similar continuities of multicomponent occupations occur at sites 7K-D-45 and 7K-D-48 on Saint Jones Neck in northern Kent County (Delaware Division of Historical and Cultural Affairs 1978). The Wilgus site (7S-K-21), which is the location of a Delmarva Adena and Carey complex macroband base camp (Custer, Stiner, and Watson 1983), and which is located in the central portion of the distribution of the Slaughter Creek complex, also has a Slaughter Creek complex occupation. Archaeologists found a large storage feature that contained stratified remains and a small quantity of artifacts, along with abundant faunal and floral remains that indicated a variety of wild plant foods. The similarity of these faunal remains to the earlier components at the same site underscores the similarity of the subsistence activities from Early/Late Woodland times.

The Island Field site (7K-F-17), in addition to being the site of a Webb complex cemetery, is also the location of a Late Woodland microband base camp or village site. No description of the Late Woodland components was ever published, although Artusy and Griffith (1975) do note the presence of a semisubterranean pithouse feature at the site. Recent excavations by the University of Delaware's department of anthropology revealed the presence of a sheet midden adjacent to the pithouse structure. Extensive flotation of the midden has recovered a large number of seed plant remains, many of which initially appear to be amaranth and chenopodium.

Microband base camps of the Slaughter Creek complex also appear as single component sites. The Indian Landing site (7S-G-1) along the north shore of Indian River Bay in Delaware is one of the best-studied of these sites (Thomas et al. 1975). Thirteen excavated storage and refuse pits revealed a variety of floral and faunal remains, which indicates a summer and fall occupation (Thomas et al. 1975; 83–84). Also at the site was a variety of tool forms. An interesting component of the tool kit is a series of bone tools that seems to be related to weaving or perhaps to the production of textiles such as nets or mats. The site is much smaller than the macroband base camps and seems to be related to a variety of food production activities. Similar sites include Warrington (7S-G-14), reported by Marine et al. (1964) and Griffith (n.d.a); Poplar Thicket (7S-G-22) reported by Griffith (n.d.b); and Bay Vista (7S-G-21). The Warrington and Poplar Thicket sites are interesting in that both contain semisubterranean pithouse features (Artusy and Griffith 1975), which documents the presence of these house forms at microband base camps as well as at larger macroband base camps. Radiocarbon dates from Poplar Thicket, Warrington, and Bay Vista range between A.D. 1100 and A.D. 1370 (Custer 1983a; appendix 2) and are correlated with the varying designs of Townsend ceramics noted by Griffith (1977). As we noted previously, these sites appear to be similar to Indian Landing, although the seasonal usages may not be the same. Nevertheless, they are all smaller social units than those seen at macroband base camps. Some small changes in internal site patterning are at the later Poplar Thicket site, dated at A.D. 1370, which looks like a more tightly defined small cluster of dwellings and features (Griffith n.d.b.). It is also possible that this site may represent a small nuclear family-based hamlet, rather than a microband base camp with a less structured social unit composition.

Other less completely studied microband base camps from the Delaware Bay and Atlantic Coast zones include the Lewes High School site (7S-D-5) reported by Omwake (1948) and others (Anonymous 1951a); the Ritter site (7S-D-2,3), which produced the only corn remains found in Delaware, reported by Omwake (1951, 1954b, 1954d); the Derrickson site (7S-D-6) reported by an unknown author (1951b); the Miller-Toms site reported by Omwake (1954c); a series of sites listed in the Maryland site files including 18WO59, 18WO45, and the Saint Martin site (18WO10) (MGS site files); the Sandy Point site, which is listed on the National Register of Historic Places (MHT 1974d), and other sites on Sinepuxent Neck reported by Omwake (1945); and a series of sites listed on the Virginia Eastern Shore Atlantic Coast that include the Fleming site (44AC8), the Bell site (44AC12), 44AC19, 44AC20, 44AC81, 44AC108-110, 44AC114, 44AC205, and 44NH1 (VRCA site files). Most of these sites contain shell middens, storage/processing features, or both.

Microband base camps are also along the floodplains of the peninsula's major drainages. Representative sites within the Nanticoke drainage include the Prickley Pear Island site (7S-H-18), which contained a number of shell-filled pit features and which produced a radiocarbon date of A.D. 1015 (Custer 1983a, appendix 2), and a series of sites along the Marshyhope Creek (Corkran and Flegel 1953; Flegel 1976). Similar sites on the Choptank include the Maloney site (18CA57) reported by Gardner and Rappleye (1980); 18DO62, and 18DO126 (MGS site files); and the Holiday Park site reported by Payne (1983). Similar sites with small middens and pit features on other drainages of the southern portion of the Maryland Eastern Shore include Reeves (18WC15), which was dug by the Lower Delmarva Chapter of the Archaeological Society of Maryland (MGS site files); Willin (18DO1) reported by Bryant et al. (1951) and Hutchinson (1967), and which is listed on the National Register of Historic Places (MHT 1974c); the Chicone No. 1 (18DO11) and the Chicone No. 2 (18DO10) sites reported by Hutchinson, Callaway, and Bryant (1964); and Stasl (18WO148), Pulpwood Landing (18WC6), Indian Knoll Landing (18WO6), Rachel (18SO117), Lower Geanquakin (18SO102), Golden Eagle Landing (18SO94), Fairmont (18SO20), 18SO77, and 18WO136 (MGS site files).

The embayed coasts of the Chesapeake Bay area are also locations for microband base camps. In most cases these sites have fairly large shell middens. Pit features are less common on these sites than at the other microband base camps described previously. However, the comparable size and tool assemblages, as well as the presence of shell middens in most of the sites, suggests a similar function and role in regional settlement systems for all of these sites. Examples include a complex of seven sites on the inland side of Eastern Neck Island in the northern portion of the study area (Thompson and Gardner 1978; Thompson 1982); the Pine Bluff site reported by Marshall (1977); a series of shell middens noted on Smith Island (Gardner et al. 1978); the Steelman site (18TA103), which was tested by Joseph McNamara of the Maryland Geological Survey in 1983; 18DO115, 18DO25, 18SO31, 18SO126, and 18DO79 (MGS site files); and a series of sites on the Chesapeake Coast of the Virginia Eastern Shore including 44AC15, 44AC134, 44AC160, 44AC165, 44NH22 (VRCA site files). In two cases (18KE246 [Thompson 1982] and 44AC160) these sites contained from one to three burials, which indicates that they were inhabited by a small social unit.

Procurement Sites

Procurement sites are the most poorly known segment of the Late Woodland settlement systems of the Delmarva Peninsula. However, there are clues to their locations from present data. A recent regional survey of the Saint Jones and Murderkill drainages (Custer and Galasso 1983), a comprehensive survey of the Saint Jones Neck area (Delaware Division of Historical and Cultural Affairs 1978), and a study of interior portions of the peninsula (Kavanagh 1979) indicate that the Late Woodland procurement sites are located on small sand ridges adjacent to poorly drained woodlands and along low-order ephemeral streams in the northern portions of the study area. In the lower Eastern Shore of Maryland, small procurement sites, which appear in the archaeological record as small lithic scatters with a few discarded tools, appear along the major drainages adjacent to floodplain swamps and marshes (e.g., Thomas 1978); along the small bays and drainages leading into the Atlantic Coast barrier island lagoons (e.g., 18WO1 and 18WO127—MGS site files); on the Chesapeake Bay coastline (e.g., 18SO69, 18SO90, and 18SO95—MGS site files); and in interior settings similar to those previously noted in northern portions of the study area (e.g., 18WO14, 18WO21, 18WO126, 18WO130, 18DO111, and 18WC9—MGS site files). On the Virginia Eastern Shore, Clark (1976) notes a series of procurment sites on the western margin of Chincoteague Bay, on the Atlantic Coast. These sites are along low-order drainages and on the bay coast. Procurement sites associated with procurement of cobble resources also appear in the Virginia portion of the study area (e.g., 44AC263 and 44NH9—VRCA site files). In general, procurement sites turn up next to good hunting, gathering, and shellfish collecting locales throughout the study area.

Discussion

The above description of the Slaughter Creek complex sites shows that there is a range of variation in settlement patterns and adaptations during the Late Woodland period in the southern two-thirds of the Delmarva Peninsula. It is not surprising that Thomas et al. (1975)

were unable to distinguish the single most accurate model of Slaughter Creek settlement patterns. Depending on the site location, any one of the three models (models 3, 4, 5—table 1) may be applicable. Adaptations may also vary through time. Luckily, the large amount of data available for many varied Slaughter Creek complex sites makes it possible to generate some preliminary statements about spatial and temporal variation in settlement patterns.

Significant spatial variation in settlement patterns and adaptations occur when various portions of the Slaughter Creek complex are compared to one another. North of the Mispillion River in Delaware there are only two Late Woodland macroband base camps, the Hughes-Willis and Island Field sites; these are significantly smaller than the large macroband base camps or villages located south of the Mispillion River. The Hughes-Willis site especially looks very much like a late Middle Woodland macroband base camp except for its high concentration of storage features. Similar patterns can be seen along most of the interior drainages. Model 4 (table 1, fig. 2) as described by Thomas et al. (1975, 64) best approximates the Slaughter Creek complex settlement pattern in this area. There is continuity with Early/Middle Woodland adaptations in these zones, which is underscored by the distribution of microband base camps and their high frequency of occurrence. The fissioning of macroband base camps (Custer 1982b, 30-32) that occurred during the Middle Woodland Webb complex apparently continued into Late Woodland times in these areas. Intensified use of storage along the major drainages may be correlated with increased sedentism and population growth; however, these processes are not strongly developed on these southern portions of the Slaughter Creek complex distributions.

Different settlement systems turn up south of the Mispillion River and in coastal areas. The coastal area between the Mispillion River and Cape Henlopen includes the largest Slaughter Creek complex macroband base camp sites. Model 5 (table 1, fig. 3), described by Thomas et al. (1975, 64), best approximates the site distributions in this area. The large size of the macroband base camps and the high density of storage features at sites such as Mispillion, Slaughter Creek, and Townsend indicate an increase in population and/or sedentism beyond the levels seen at Early/Middle Woodland sites. Similarly, the location of these large sites at the interface of the middrainage settings and the Delaware coastal zone represents a break from Early/Middle Woodland patterns.

The area from Cape Henlopen south along the Atlantic Coast to the southern tip of the peninsula and the Chesapeake Bay coastal zone from the mouth of the Chester River south to the southern tip of

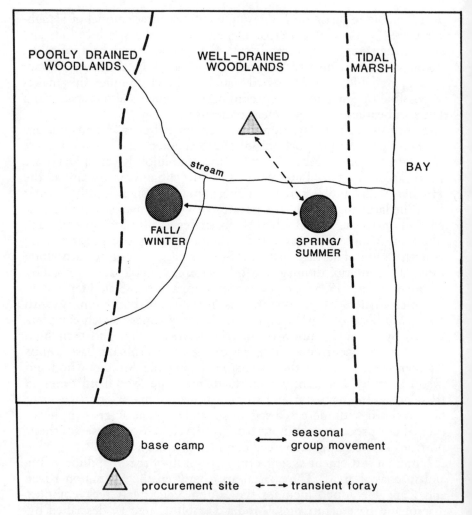

Fig. 2. Late Woodland settlement model no. 4.

the peninsula are similar to areas north of the Mispillion River in that few macroband base camps are noted in this area. The wide range of microband base camps in this area indicates that Model 3 (table 1, fig. 4), as noted by Thomas et al. (1975, 63), best applies to the Slaughter Creek complex south of Cape Henlopen and along the Chesapeake Bay coast. Also evident in these coastal areas is a high degree of continuity in settlement patterns from Middle Woodland, Late Carey, Carey, and Webb complexes.

Subsistence Systems

The shell middens and pit features of Late Woodland Slaughter Creek complex sites on the Delmarva Peninsula contain abundant ecofacts of various types that reveal much about subsistence systems. However, most of the excavated data come from the northern reaches of the study area in Delaware. This section first considers the types of

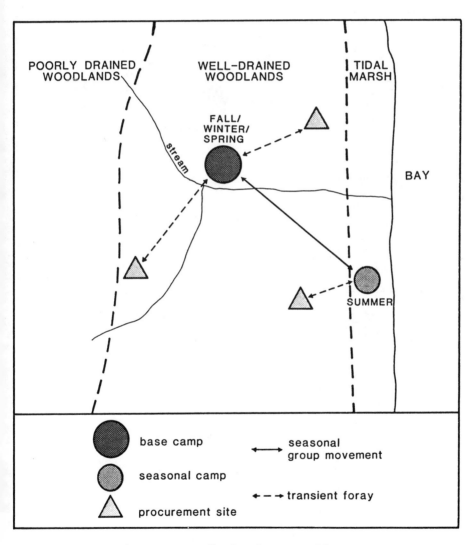

Fig. 3. Late Woodland settlement model no. 5.

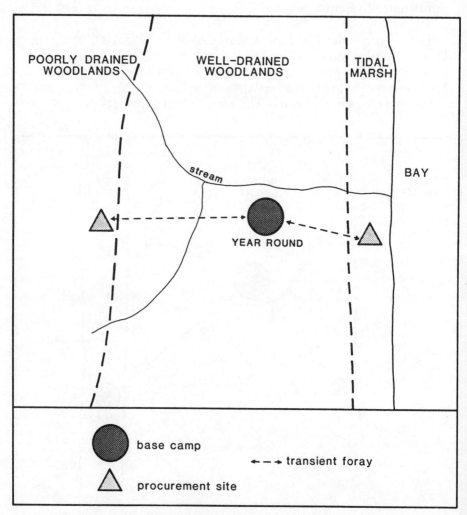

Fig. 4. Late Woodland settlement model no. 3.

food remains found at the Slaughter Creek complex sites and then discusses the role of storage features in Late Woodland subsistence systems.

Analysis of Food Remains

Because animal bone remains are easily visible in, and easily recoverable from, the archaeological record, faunal remains are the most common food remains discussed in literature. In table 2 are the species represented at five Slaughter Creek complex macroband base camps and two microband base camps. Recovery and analytical techniques have varied greatly from the earliest excavations (Slaughter Creek) to the latest (Wilgus); therefore, comparison is difficult. Nonetheless, some trends can be noted. All sites have deer remains, and where quantitative information is available, deer represents the

Table 2
Faunal Remains: Slaughter Creek Complex Sites

	sf	dr	br	rab	sq	musk	turt	tur	bds	fish	dog	Reference
Macroband base camps:												
Slaughter Creek (7S-C-1)	X	X	X	X	X							Davidson 1935
Townsend (7S-G-2)	X	X		X	X	X	X	X	X	X	X	Omwake and Stewart 1963
Mispillion (7S-A-1)	X	X		X	X	X	X					Thomas and Warren 1966b
Hughes-Willis (7K-D-21)		X				X	X					Thomas et al., 1975
Lankford (18DO43)	X	X		X	X	X	X	X	X	X	X	Flegel 1975a, 1975b, 1976
Microband base camps:												
Indian Landing (7S-G-1)	X	X		X	X	X	X	X	X			Thomas et al., 1975
Wilgus (7S-K-21)	X	X		X				X	X		X	Custer et al., 1983

X = Present
sf = shellfish
dr = deer
br = bear
sq = squirrel
musk = muskrat
turt = turtle
tur = turkey
bds = birds

major food source in conjunction with shellfish. The other faunal remains listed in table 2 seemingly represent a small portion of Late Woodland diets. At the sites where seasonal indicators are available (Hughes-Willis, Indian Landing, and Wilgus), deer is a major food source during all seasons of the year. Note too that deer is the major meat source at both macro- and microband base camps. Unfortunately, there are insufficient quantitative data to determine if the proportional use of deer varies from season to season among various site types. Finally, the sites that produced these data are all from the Delaware Bay coastal zone and Atlantic Coast, except for the Hughes-Willis site. Based on the Hughes-Willis data, we can also tentatively hypothesize that similar patterns of faunal resource utilization will be seen at interior major drainage macroband base camps. As far as the Chesapeake Bay region is concerned, deer bone occurs in a few northern coastal Slaughter Creek complex shell middens tested by Wilke and Thompson (Custer and Doms 1983), but no further interpretation is possible.

Discussion of plant food resources is even more limited than that of faunal resource utilization because flotation techniques that recover small seeds and plant fragments have only recently been applied to sites from the Delmarva Peninsula. Larger plant remains, such as hickory nut shells or charred corncobs, are sometimes noted in older reports; nonetheless, the study of a range of plant foods is restricted primarily to recently excavated sites. The recently collected information on Slaughter Creek complex plant food utilization appears in Table 3. One striking feature is the absence of cultigens, especially corn. In fact, the only corn noted from a Delmarva Peninsula Late Woodland site is from Ritter (7S-D-2,3), a microband base camp reported by Omwake (1951, 1954b, 1954d). Given the fact that artifacts smaller than corncobs were found at some of the older excavated sites, one can reasonably assume that if corn were present at other sites, it would have been found. However, smaller cob fragments and kernals may have been missed.

All of the sites listed in table 3 were subjected to flotation, so their faunal remains are comparable. Hickory nuts were in all but one sample and were present in sufficient quantities at the Hughes-Willis site for Thomas et al. (1975, 77) to suggest that their processing was a major activity at the site. Amaranth and chenopodium, which produced edible seeds (Moeller 1975), are present in very large quantities in three samples, and along with hickory nuts they represent up to 90 percent of the floral resource assemblages. Other floral resources are minor components of the assemblages and, presumably, Late Woodland diets. Note that amaranth and chenopodium are available during late summer and early fall, while hickory is primarily available in the fall. At the Hughes-Willis site, the fall availability of other species is indicated within the refuse (Thomas et al. 1975, 77). However, at the Island Field and Wilgus sites, the seed and nut contents of middens and pit features were not congruent with seasonal indicators of faunal remains and shellfish (see Custer et al. 1983). These incongruities indicate that storage of seed and nut foods may have been part of Slaughter Creek complex subsistence systems. The role of storage is discussed in more detail in the following section.

The importance of shellfish resources has been stressed in many studies of the Slaughter Creek complex (see Thomas 1973). However, until recently, there were no methods available to assess the role of shellfish resources in prehistoric subsistence systems. Recent work by Brett Kent (1982) has provided researchers with numerous ways to

Table 3
Floral Remains: Slaughter Creek Complex Sites

Site	Corn	amar	chen	hick	eel	hack	black	lot	Reference
Wilgus (7S-K-21)		X	X	X	X	X			Custer et al. 1983
Island Field (7K-F-17) sheet midden		X	X						Doms and Custer 1983
Island Field (7K-F-17) pit feature		X	X	X					ibid.
Indian Landing (7S-G-1)				X					Thomas et al. 1975
Hughes-Willis (7S-D-21)				X			X	X	ibid.

X = Present

amar = amaranth
chen = chenopodium
hick = hickory
eel = eelgrass

hack = hackberry
black = blackhaw
lot = lotus

assess important features of the oysters utilized by Slaughter Creek complex groups, including season of death, the type of environment in which the oysters lived, and attributes of their surrounding environment. Although shellfish resources were at many Late Woodland sites, systematic collection and notation of species is only a very recent research activity. Nonetheless, examination of earlier site reports, as well as more recent reports, shows two main contexts for shellfish remains: middens (either stratified or sheet) and as fill in refuse pits. In many cases the shells have been burned, which indicates roasting as a possible means of preparation. Also, recent analyses from middens at the Island Field and Wilgus sites show few valves with broken ends where they had been opened, which further supports the view that the shellfish had been either roasted or steamed.

Slaughter Creek complex sites have a wide variety of shellfish species including oyster, hard and soft clam, ribbed mussel, whelk, and periwinkle (Thomas et al. 1975, 84; Custer at al. 1983; Omwake and Stewart 1963, 6). However, oysters and clams generally account for 90–95 percent of the shellfish remains. To date, there has been no thin-section analysis for season of death on clams from any Slaughter Creek sites; however, Kent's (1982) techniques of oyster analysis have been applied to oyster shell from the Wilgus and Island Field sites. At the Wilgus site, oyster shell comprised the majority of shell fill in a large (2.5m diameter) pit and most were mature specimens that were collected from an intertidal mudflat during very late summer/very early spring (Custer et al. 1983). The oysters were fairly large, but their nutritive value would have been low (Kent n.d.). The oysters analyzed from the Island Field site came from a small bushelbasket-sized pit and are all "coon oysters," ones that are long and thin and grow in dense clusters in intertidal zones. The estimated four hundred total valves in the feature probably represent a single cluster and the oysters, although old—some up to nine years—are quite small (Doms and Custer 1983). These oysters were also harvested during late winter/early spring and would have had exceptionally low food value (Brett Kent, personal communication). Since similar assessments of shell from other earlier Early/Middle Woodland sites (McNamara 1982b; Custer et al. 1983) show similar patterns of seasonal use, we suggest here that one major role of shellfish, particularly oysters, was as a "survival food" that was consumed during times of dietary stress. Analyses of hunter-gatherer subsistence systems show that the early spring is one such period of stress (Jochim 1976). Note that this usage does conform to the predictions generated by the ecological models used by Thomas et al. (1975) but does not fit with the predictions of "optimum foraging theory"-based models (Cameron 1976; Gilsen

1979, 1980). Further work at larger sites with large-shell samples is necessary to fully understand the role of shellfish resources in Late Woodland subsistence systems.

Storage

The archaeological record of Slaughter Creek complex sites contains large numbers of engimatic pit features. Certainly at least one of their functions was to hold refuse because the pits are almost always filled with shell, artifacts, and various types of food remains. However, careful excavations of these features suggest that they originally had other functions. In some cases pits showed evidence of burning including reddened soil, presence of fire-cracked rock in the feature fill, and burned shell (Omwake and Stewart 1963; Davidson 1935a, 1935b; Zukerman 1979a, 1979b). These data suggest that some pits may have served as roasting ovens with direct firing and/or hot-rock heating.

Other features have different attributes that suggest other original uses. The Warrington site has pit floors lined with carefully placed shells (Griffith n.d.a.), as well as Bay Vista, Townsend (Omwake and Stewart 1963; unpublished excavation notes at Island Field Museum), and numerous sites on the Sinepuxent Neck (Omwake 1945). We suggest that these pits may have been used for storage of foodstuffs with the shells improving drainage within the pit. Prepared sand floors are also present in some features (e.g., Wilgus site—Custer et al. 1983), which suggests a similar function. Even more convincing evidence comes from the Poplar Thicket site (Griffith n.d.b.) where a pit feature appeared to have a lining of wickerwork textile and contained a large mortar and several grinding stones. Because grinding stones would most likely have played a role in seed-plant processing, these resources may have been stored in, or processed near, these features.

The best indication of storage use comes from a consideration of pit contents. As was noted earlier, analysis of pit feature contents at the Island Field and Wilgus sites revealed late summer fall seed and nut resources in association with oyster shells and faunal remains indicating a winter/spring deposition (Doms and Custer 1983; Custer et al. 1983). This incongruence shows that seeds and nuts were the items stored in the pit features. Combined with shellfish and deer, these stored food resources carried groups through the lean times of early spring and allowed a relatively stable and sedentary lifestyle in some areas. Based on current data, this interpretation most likely applies to sites along the Delaware Bay Coast and in the Atlantic coastal zone.

Social Organizations and Complexity

Inferences about Late Woodland social organizations and relative social complexity can be made based on two classes of data: archaeological and ethnohistoric data.

Inferences from Archaeological Data

One source of inferences about social organizations and complexity is the analysis of settlement patterns. In most cases, the arrangements of various base camps and procurement sites within settlement pattern models (figs. 2–4) suggest an egalitarian fusion-fission cycle of social organization with only a limited development of residential stability. In other areas, specifically the Delaware Bay coastal zone and the northern cases of the Atlantic coastal zone, the presence of very large sites, such as Slaughter Creek, Mispillion, and Townsend, with high densities of pit features and houses, suggests a greater residential stability. This higher degree of residential stability may also correlate with higher population densities and slightly more complex social organizations. The Slaughter Creek site, with its main macroband base camp and possible associated hamlets, also may indicate the development of a residential site hierarchy that can be linked to the initial appearance of ranked societies (Carniero 1981; Fried 1967). However, further research at the Slaughter Creek site is necessary to develop these hypotheses.

The presence of multiple burial features at some of the larger sites, such as Slaughter Creek, Townsend, and Thompson's Island, may also indicate the supralocal organization of communities in that several communities may have cooperated to sponsor the community organization of burial rituals. These organizations can be associated with incipient ranked societies (Carniero 1981; Price 1981, 1982). Although these multiple burials are much smaller than true ossuaries, as Ubelaker notes (1974), the presence of evidence for exhumation and secondary burial treatments of individuals at some Slaughter Creek complex multiple burials (Thomas 1973) suggests supracommunity organizations. Similar burial treatments occur among some ethnohistorically documented coastal Algonkian groups (Feest 1973; R. Flannery 1939) that had some characteristics of ranked societies.

Although there are some indications that the Late Woodland Slaughter Creek complex groups had some attributes of incipient ranked societies, sites with these attributes generally restrict themselves, to northern coastal areas, although it is possible that some more complex sites may have existed on the Lower Choptank drainage (Custer 1983c; Weslager 1942). Also, a comparison with earlier Early/Middle Woodland social organization places the Late Woodland organizations within a temporal perspective. Late Archaic/Early Woodland Barkers Landing and Delmarva Adena complexes, as well as late Middle Woodland Webb complex societies, have been characterized as incipient ranked societies with Delmarva Adena groups probably the most complex (Gardner 1982, 72–77; Custer 1982b, 34–36; Custer 1983a; chap. 4). The prime indicators of these complex societies are burial ceremonialism, supralocal exchange in exotic raw materials and sumptuary goods, and settlement hierarchies. Comparisons between these earlier sites' attributes and those of Slaughter Creek sites show marked differences. First, even though the Late Woodland multiple burials hint at some supracommunity organizations, they lack the differential distributions of grave goods that characterize earlier cemeteries and that provide the clearest indications of differential social statuses. In fact, the Late Woodland burials, including the isolated occurrences of single burials, usually lack any kind of grave goods. Second, by the beginning of Late Woodland times, all indications of long distance trade and exchange disappear from the Delmarva Peninsula. Finally, the Delmarva Adena settlement pattern of isolated complex cemeteries surrounded by separate support sites is not seen in Late Woodland times when multiple burial features are directly associated with the largest sites. Thus, although a few of the larger sites of the Slaughter Creek complex may have some attributes that suggest greater social complexity than that of groups at other sites, these Late Woodland societies would still be less complex than their Early/Middle Woodland predecessors.

Changes in social organizations and group interaction patterns can also be noted at different times during the Late Woodland period. During initial Late Woodland times, Townsend ceramics of the Slaughter Creek complex show marked similarities to Minguannan, Overpeck, and Bowmans Brook ceramics of the Delaware River Valley (Lopez 1961; Griffith and Custer n.d.). By ca. A.D. 1300 (Griffith 1977, 148) there seems to be a change in ceramic styles in southern Delaware with corded horizontal motifs, similar to designs found on Potomac Creek ceramics of the western shore of the Chesapeake Bay (Stephenson 1963), becoming more common. Hughes (1980, 205–14) notes similar patterns on the Lower Eastern Shore of Maryland. At the same time, there seems to be a southward shift of populations using Townsend ceramics (Griffith 1977, 145–50). Accompanying this southward shift of populations is the appearance of nonlocal Woodland II ceramics such as Potomac Creek and Keyser Farm varieties at the Robbins Farm site in central Kent County

(Stocum 1977). The location of the appearance of these nonlocal ceramics is especially interesting in that they are found at the southern margin of the area of Delaware that seems to have a low population density during Late Woodland times. This possible low-population zone extends from the Modern Chesapeake and Delaware Canal to at least the headwaters of the Saint Jones drainage. The apparent southward shift of the Slaughter Creek complex populations could have extended this zone as far south as the Mispillon drainage by A.D. 1300. The presence of nonlocal ceramics may represent an initial expansion of nonlocal populations using markedly different ceramics into this low population density zone. Gardner and Carbone (n.d.) note that numerous population disruptions are apparent throughout the Late Woodland Period in the Middle Atlantic and the appearance of Potomac Creek ceramics in central Delaware may be correlated with one of the later population disruptions in the lower Potomac River Valley. Future fieldwork should be able to test the validity of this hypothesis.

Ethnohistoric Data

Ethnohistoric observations support the interpretation of the archaeological data noted above. We do not give a detailed analysis of the ethnohistoric data here, but offer only a few observations. For a more complete assessment of the ethnohistoric and historic sources, we refer the reader to critical bibliographies compiled by Weslager (1978) and Porter (1979), as well as Weslager's (1972, 1983) studies of the Delaware and the Nanticoke.

While it is clear that most of the Contact-era societies in the study area were Algonkian speakers (Goddard 1978b) and can be classified within the central coastal Algonkian culture area (Flannery 1939), Feest notes that the ethnohistoric record shows a complicated variety of ethnic groups and political units on the Delmarva Peninsula at the time of the initial European Contact. Throughout the seventeenth century, political alliances shifted among recognizable ethnic groups such as the varied alliances among the Choptanks, Assateagues, and Nanticokes (Feest 1978b, 242). This factor complicates the identification of specific attributes of individual ethnic groups that may be identified with archaeological cultures. Also, the rapid disruption of Indian societies in the face of European Contact complicates the problem of identifying pristine social organizations and characterizing their relative complexity. Even where there is an abundance of historic and archaeological evidence, such as the Delaware Bay coastal zone in the northern portions of the study area, there is debate over the identification of ethnic boundaries. For example, Weslager (1983) and Goddard (1978a, fig. 2) believe that Delaware (Unami) groups inhabited the entire Delaware Bay coast south to Cape Henlopen. Others think that Cape Henlopen groups represented a social unit distinct from Unami groups, with a markedly different adaptation from groups in the upper Delmarva Peninsula. Whatever the case for the Cape Henlopen groups, the ethnohistoric record does document a variety of localized ethnic groups with shifting alliances at initial Contact times.

In the ethnohistoric record are a variety of statements that can be related to relative sociocultural complexity. Beginning in the southern portion of the study area, Turner (1982b, 48) notes that Smith (Arber 1910, 26, 55) and Strachey (1853, 49) present statements that two "districts" on the Virginia Eastern Shore, Accomac and Accohannoc, were under the rule of Powhatan, whose main village was located on the western shore of the Chesapeake Bay, and that they paid tribute to him. However, even though they were tributary, the districts did *not* participate in Opechancanough's war against the English in 1622 (Turner 1982b, 48; Beverly 1947, 51; Neill 1968, 366). Population estimates for these two districts developed by Turner from Smith's (Arber 1910) figures are 340 people (80 warriors) in the Accomac district and 170 people (40 warriors) in the Accohanoc district. Both population estimates could apply to either a single large community or a series of separate, localized social units that acted together in dealing with the more complex, and more powerful, groups of the Virginia Tidewater.

Feest's (1978b) summary of ethnohistorical information from the central Chesapeake Bay area includes numerous references to groups living on the Delmarva Peninsula between the Choptank River and the modern Maryland-Virginia border. Within this region, several groups (Assateague, Accomac, Nanticokes, Pocomokes) have "chiefs", or paramount leaders with varying degrees of centralized authority (Feest 1978b, 240–41). The Choptanks had the least centralized organization and many local leaders are noted in deeds and court records (Feest 1978b, 242). Post-1650 data indicate that there was a series of upper-level social statuses inherited along the patrilineal lines (Feest 1978b, 245). However, these authority structures may have existed during pre-Contact times and developed in response to increased interactions with Europeans and/or more complex groups of the Virginia Tidewater. It is also possible that Europeans superimposed their own conceptions of how centralized-authority structures should work upon groups who never had any such organizations (Jennings 1975).

The ethnohistoric records also provide some information on burial customs that can be related to questions about social organizations and social complexity. Archaeologists have noted secondary burial treatments and ritual reburial in multiple interments for both commoners and high-status individuals among a number of groups on the southern Delmarva Peninsula (Heckewelder 1819; Zeisberger 1910; Feest 1973). The mortuary practices they described would correspond to some of the multiple burials seen in the Late Woodland archaeological record, as well as in the archaeological record of refugee groups of the Contact area (Kinsey and Custer 1982). There are also some hints at special burial treatments of high-ranking individuals. Feest (1973, 6) notes Browne's (1887, 481, 483) description of an Assateague chief's burial site, which contained numerous perishable grave goods, and Speck's (1927, 41) analysis of historic documents that describes the ritual treatment of the remains of the local chief Wynicaco within a charnel (quioccosan) structure (see also Weslager 1942). These descriptions would seem to indicate some special burial treatment of individuals of high status, but this special treatment may not be visible archaeologically.

A final topic for discussion that appears in certain historical records is the description of Indian cornfields in the southern portion of the Delmarva Peninsula. Marye's (1936a, 1936b, 1937, 1938, 1939, 1940) analyses of land sales and old maps from the 17th and 18th centuries record numerous references to "Indian fields." However, these instances of aboriginal agriculture were very late phenomena that occurred as a response to the effects of European Contact (Ceci, 1977).

In sum, the data from the ethnohistoric literature roughly corresponds to the inference from the Late Woodland archaeological record for the southern Delmarva Peninsula. The ethnohistoric record does seem to indicate that there were supralocal multicommunity social organizations controlled by limited numbers of high-status individuals. These kinds of organizations are the minimal attributes of chiefdoms and the most simple types of ranked societies (Carniero 1981). However, these multicommunity alliances are relatively ephemeral and are not found in all portions of the study area. We also suggest that these organizations may have developed in response to pressures from groups on the western shore of the Chesapeake Bay. The tributary status of the Accomac and the Accohanoc areas within the Powhatan Confederacy at Contact clearly documents the existence of these pressures. Nonetheless, even if these multicommunity organizations were not late developments, they were small enough to represent only the most simple chiefdoms similar to those described by Oberg (1955) for lowland South America and Steward (1948) for the intermediate area and the Caribbean. Furthermore, these societies also lacked the long-distance trade and exchange and elaborate mortuary ceremonialism that characterized their more complex Early/Middle Woodland predecessors. More important, the simple ranked societies of the southern Delmarva Peninsula lacked the redistributive economies that were based partly on agricultural surpluses characterizing the petty chiefdoms of the western shore of the Chesapeake Bay (see Turner's paper in this volume and 1982a, 1982b, 1976; Potter 1982).

Conclusion

A variety of sociocultural adaptations characterized the central and southern Delmarva Peninsula during Late Woodland times. Continuities between late Middle Woodland and Late Woodland adaptations are apparent in some areas, but significantly evolutionary changes in adaptations crop up in other places. The clearest indications of these changes come from archaeological sites in the Delaware Bay coastal zone in the northern portions of the study area. Significant features of these adaptation changes are intensified use of stored resources, particularly plant resources, and shellfish resources.

These trends can be related to local ecological conditions. The drainages of this area between Cape Henlopen and the Mispillion River are shorter and somewhat truncated, compared to the drainages to the north and to the south. Thus, effects of sea level rise would be more pronounced and the size of the productive resource zone smaller. In these marginal settings, groups may have been forced to intensify food procurement, begin to use storage, and possibly even to add agricultural food production to their subsistence base. These options could have been taken during years of low resource productivity that may have been caused by the oscillations of temperature and moisture noted for the recent prehistoric past by Carbone (1982), or they may have been undertaken because local population densities prevented the movement of groups to more productive zones (see Binford 1983, 204–13). Whatever the case, as these changes were made, population could grow and groups could become increasingly sedentary (Catlin et al. 1982). Similarly, population growth and sedentism would create a greater need for increased use of storage and agricultural food production systems. In spite of the high local population densities, which seem to be even greater than levels recorded during Early and Middle Woodland times, there is little evidence of ranked societies such as would be shown by elaborate burials or exchange networks. Although disarticulated burials and ossuarylike features occur at some of the larger Slaughter Creek complex sites, they are too small to be clearly related to attributes of ranked societies (Ubelaker 1974), as are the true, large ossuaries associated with petty chiefdoms in the lower Potomac River Valley (Potter 1982). As such, these sites appear to represent villages of relatively constant egalitarian social group composition with a few indications of simple multicommunity organizations. Interregional interaction among social units through exchange seems to be minimal, although there is enough intraregional interaction to produce similar ceramic design motifs throughout the Slaughter Creek complex sites. Especially interesting is the fact that these large sites seem to appear in zones that do not have the highest natural productivity. However, the appearance of agricultural food production systems in relatively marginal environments frequently occurs throughout the world. In southern Delaware, the environments of the truncated drainages at first may be sufficiently rich to provide a biosocial environmental setting that favors some intensification of food production; however, the necessary social environments for the development of more complex ranked societies do not seem to be present, and extensive trade and exchange and mortuary complexes, such as those occurring during the Early/Middle Woodland period, did not develop. The crucial variation in the social environments between Early/Middle Woodland and Late Woodland groups may be related to the fact that Early/Middle Woodland groups were circumscribed in highly productive environments where intensification of food production generated temporary surpluses that could be stored by some people and "invested" in either status symbols and group relationships that could be exchanged or in the sponsorship of ritual events that verified the high status of the participating individuals (Harris 1979). In contrast, the Late Woodland groups of the lower Delaware Bay may not have been sufficiently circumscribed to force the generation of a true surplus, even with agricultural food production, that could be used for anything beyond normal subsistence needs. Therefore, since there was nothing to "invest" in the accoutrements of ranked societies, more complex social organizations never developed. It is also possible that in northern portions of the study area the natural environment was not rich enough or the subsistence systems insufficiently intensified to support the population growth that allows circumscription to "cause" increases in sociocultural complexity.

In short, environmental productivity in the absence of agriculture and the degree of population growth and consequent circumscription are the critical variables in this hypothetical model of variable sociocultural complexity during Late Woodland times on the Lower Delmarva Peninsula. Future research in testing these explanations should help to increase the understanding of general cultural evolutionary processes.

3

Late Woodland Cultures of the Middle and Lower Delaware River Valley and the Upper Delmarva Peninsula

R. MICHAEL STEWART
CHRIS C. HUMMER
JAY F. CUSTER

Late Woodland cultures of the Middle and Lower Delaware River Valley and the Upper Peninsula show a wide range of settlement/subsistence systems and varying degrees of social complexity. The societies of these areas may all be equated with protohistoric, Unami-speaking Delaware Indians (Goddard 1978). This paper presents three crosssections of the different environmental zones and related adaptations associated with the valley: the Piedmont Uplands and the Upper Delmarva Peninsula; the Coastal Plain/Piedmont transition near the Fall Line at Trenton; and Piedmont, Highlands, and Ridge and Valley sections of the Middle Delaware Valley. The Late Woodland environments of these zones were essentially those of historic times. Minor climatic perturbations may have modified the nature and extent of "open" types of habitats during the period (Carbone 1976). Continual sea level rise would have gradually altered the extent of tidal influence throughout Late Woodland times (Kraft and John 1978).

As used here, the term *Late Woodland* refers solely to a chronological period circa A.D. 900/1000 to 1620 or the time of European

---

For helpful comments the authors would like to thank Marshall Becker, John Cavallo, Debbie Fimbel, Dan Griffith, Kurt Kalb, F. Dayton Staats, and Fred Kinsey.

Much of the research involving the Trenton area and the Abbot Farm was funded by the Federal Highway Administration and the New Jersey Department of Transportation as part of a cultural resource survey performed by Louis Berger and Associates in advance of interstate routes 195 and 295 and New Jersey routes 29 and 129. The authors greatly appreciate these organizations' continued support and interest in regional archaeology. The Delaware Division of Historical and Cultural Affairs and Maryland Historical Trust also supported significant research. The Archaeological Society of New Jersey provided funds for two radiocarbon dates at the Williamson site.

Contact. This time period broadly corresponds with the regional use of triangular projectile points and the incidence of complex ceramic design motifs. A common cultural adaptation is not part of this definition, as will be evident in the following discussions. The treatment afforded each crosssection of the Delaware Valley varies according to the extent of previous archaeological research. Basic features discussed include settlement and subsistence patterns, intrasite patterning or structure, and the complexity of social organization. The paper concludes with a brief examination of some of the processes that might have been responsible for the cultural variation observed.

Piedmont Uplands and the Upper Delmarva Peninsula

Culture Area and Cultural Complexes

The Piedmont Uplands and Upper Delmarva Peninsula, as illustrated in figure 5, correspond to the distribution of the Minguannan complex as defined by Custer (1983). The Minguannan complex, dated to between A.D. 1000 and A.D. 1600, includes diagnostic artifacts such as triangular projective points and Minguannan ceramics, which have been described in other publications (Custer n.d.; Griffith and Custer n.d.). Minguannan ceramics are characterized primarily by broadline-incised geometric designs and are sand- and/or grit-tempered with smooth surfaces. Minguannan ceramic designs are virtually identical to those seen on Townsend ceramics (Griffith 1982) and are also very similar to Moyaone (Stephenson 1963), Overpeck (Forks of the Delaware Chapter 1980), and Bowmans Brook ceramics (C. Smith 1950) of adjacent areas (Griffith and Custer n.d.).

Settlement pattern similarities throughout the Upper Delmarva Peninsula and Piedmont Uplands also define the Minguannan complex (Custer 1983). Minguannan complex settlement patterns have a series of habitation sites, described as macroband base camps, that do not contain house patterns, storage features, or defined midden areas. There are no indications that these sites represent sedentary villages, or even farmsteads, similar to those seen in adjacent regions during Late Woodland times. All known Minguannan macroband base camp locations are the same as earlier Late Archaic through Middle Woodland macroband base camps. Smaller base camps, presumably used by smaller social units, and procurement sites also characterize the Minguannan complex; however, their distributions are poorly known

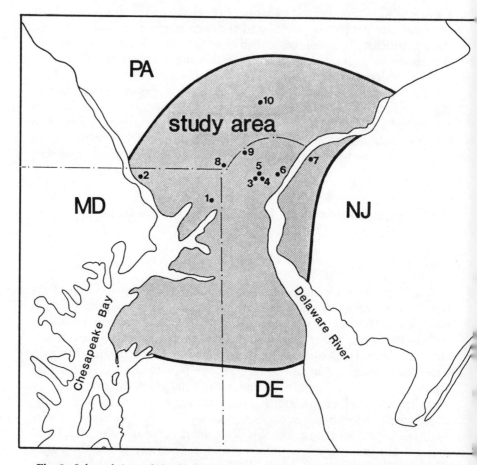

Fig. 5. Selected sites of the Piedmont Uplands and Upper Delmarva Peninsula. 1: Hollingsworth; 2: Conowingo; 3: Delaware Park; 4: Clyde Farm; 5: Julian Powerline; 6: Crane Hook; 7: Raccoon Point; 8: Minguannan; 9: Mitchell Farm; 10: Webb.

and they do not clearly define differences between the Minguannan complex and adjacent Late Woodland complexes.

Settlement Pattern Data

Analysis of the Minguannan complex settlement pattern data utilizes the site typology developed by Gardner (1982) for the analysis of Middle Atlantic Woodland settlement patterns. Gardner's typology recognizes a basic distinction between habitation sites and specialized procurement sites. Habitation sites are further divided into sites with large numbers of social units and individuals and sites with fewer people and social units. Distributions of tool types and quantity of remains are the major factors used to distinguish different site types. This typology has been successfully applied in the local area (Custer 1983; Custer and Wallace 1982). The distribution of each type of site will be described for each of the major physiographic divisions of the Piedmont Uplands/Upper Delmarva Peninsula, which includes Coastal Plain, Fall Line, Interior Swamp, and Piedmont Uplands.

Late Woodland macroband base camps throughout the study show little change in location from preceding Late Archaic through Middle Woodland macroband base camps. In the Coastal Plain there are numerous examples, including Arrowhead Farm (Wilke and Thompson 1976) and Herring Island (Henry Ward, personal communication) in Maryland. At Herring Island there is a slight shift in settlement location within the island itself; however, the site locations do show considerable overlap. Controlled surface collections from Arrowhead Farm indicate the presence of some specialized activity areas with distributions similar to those seen at Late Archaic through Middle Woodland sites. Both sites are located adjacent to brackish sections of the Chesapeake Bay and its tributaries. In Delaware, the Hell Island site (Thomas 1966a; Galasso 1981) shows almost complete overlap of Late and Middle Woodland components and is associated with extensive salt marshes. Late Woodland macroband base camps in New Jersey are also associated with salt marsh settings and include the Indian head site (Mounier 1975); Raccoon Point (Kier and Calverly 1957); and Woodbury Annex (Cross 1941, 67–71). All of these sites show considerable overlap of Middle and Late Woodland components.

The Fall Line area has two macroband base camps. The Hollingsworth Farm site (Thomas 1982) and the Delaware Park site (Thomas 1981) are located at fresh and brackish water interfaces on the Elk River and White Clay Creek, respectively, and contain substantial Middle Woodland components in addition to the Late Woodland materials. The large interior swamps in the vicinity of Wilmington, Delaware, are a special environmental setting within the Fall Line area (Custer 1983) and are the foci of three large Late Woodland macroband base camps. The Crane Hook site (Swientochowski and Weslager 1942), the largest of the three sites, was destroyed by construction of the Wilmington Marine Terminal. There are extensive Middle and Late Woodland materials in the existing collections from the site. A similar situation exists for the Newport site (Custer 1982a), which is located on the lower Christina River. The third site from this area, Clyde Farm (Custer 1982a), is still relatively intact and shows extensive overlap of Middle and Late Woodland components; however, there may be a slight upstream shift of Late Woodland settlements along the lower reaches of the White Clay Creek dating from the early period of European Contact (Custer 1982a).

In the Piedmont Uplands there are four macroband base camps of the Late Woodland Minguannan complex. The Mitchell Farm site (Custer 1981), located in the Hockessin lowlands of northern Delaware, is associated with a series of sinkholes and shows a complete overlap of Late Archaic through Late Woodland materials. Two sites located in the floodplains of the minor drainages of the region, the Minguannan (Wilkins 1976) and the Webb site (Custer n.d.) in Chester County, Pennsylvania, were extensively excavated and show significant overlaps of Late Archaic through Late Woodland materials. At the Webb site, controlled surface collections and excavations show a slight shift of habitation locations. From late Archaic through Middle Woodland times the major occupations at the site are on a small knoll adjacent to a poorly drained floodchute of the White Clay Creek. Late Woodland occupations overlap with this area; however, the majority of the Late Woodland ceramics from the site are from another small knoll 30 meters back from the floodplain (Custer and Wallace 1982, 161). Also significant in the Piedmont Uplands is the Conowingo site (Stearns 1943; McNamara 1982a, 1983a) located in Maryland at the confluence of the Susquehanna River and Octoraro Creek. Artifact distributions and ceramics at Conowingo are more similar to the Minguannan complex than to the Shenks Ferry and Susquehannock cultures of the Lower Susquehanna River Valley. Although much of the site was destroyed by construction of a power-generating dam, early research by Stearns (1943), more recent excavations by McNamara (1982a, 1983a), and analyses of personal collections (John Witthoft, personal communication) indicate artifact patterns more typical of macroband base camps than sedentary villages. McNamara's excavations also show some overlap of Late Archaic through Late Woodland artifact distributions.

Information on microband base camps is not as extensive as that for macroband base camps. No microband base camps have been identified in the Piedmont Uplands and only four possible microband base camps are known in the three remaining physiographic zones. Two sites from the Coastal Plain, the Hessian Run site in New Jersey (Mounier 1979) and the Upland Victorian site in Delaware (Kevin Cunningham, personal communication), were identified during Department of Transportation surveys and have not been extensively studied. It is not clear whether earlier components are associated with Late Woodland remains at these sites. The Green Valley site complex (Custer, et al. 1981) in the Delaware Fall Line zone includes a series of microband base camps associated with cobble beds and shows considerable overlap of Late Archaic through Late Woodland deposits. The Woods site (Faye Stocum, personal communication) in the Interior Swamp section of Delaware represents a microband base camp and is a part of a series of disturbed sites (C. A. Weslager, personal communication) that shows a slight shift up the Christina River drainage from Late Archaic through Late Woodland times. This shift is similar to that noted for the Clyde Farm location and is probably associated with alterations of the fresh and brackish water interface resulting from sea level rise. Comparable adjustments in settlement locations have been noted in other coastal settings in Delaware and indicate continuity in adaptations during Late Archaic–Late Woodland times (Custer 1982b, 1983).

Data on procurement sites are even more scanty than that for microband base camps. No information is currently available for the Fall Line and Interior Swamp sections. In the Delaware Coastal Plain four sites (7NC-D-70, 75, and 7NC-E-43, 45) are examples of Late Woodland procurement sites (Custer, Catts, and Bachman 1982; Custer and Bachman 1982). These sites contain Late Archaic through Late Woodland materials and show a common association with poorly drained woodlands and the head of small streams. Numerous examples of procurement sites have also been noted in the Piedmont Uplands (Custer and Wallace 1982). These sites tend to be located on the edges of rolling knolls adjacent to ephemeral streams and are composed primarily of small scatters of tools, points, bifaces, and limited amounts of debitage (Custer and Wallace 1982). In many cases, Late Archaic through Late Woodland projectile points have been found at the same procurement site (Custer 1980; Kinsey and Custer 1982).

In short, overall settlement pattern is based upon large macroband base camps that are located in the most productive settings such as brackish water marshes, floodplains of major streams, and poorly trained sinkhole areas. These settings are also the locations of Late Archaic through Middle Woodland macroband base camps. These very similar land-use patterns show continuity of adaptation. There is similar continuity for microband base camps and procurement sites. The similarities in all components of the settlement systems of Middle Woodland and Late Woodland times further underscore continuity in adaptations. Additional indications of continuity come from analysis of excavation data from specific sites.

Site-Specific Data

Excavation data are available for some of the macroband base camps including Hell Island (Thomas 1966a; Galasso 1981), Delaware Park (Thomas 1981), Crane Hook (Swientochowski and Weslager 1942), Newport (Custer 1982a), Clyde Farm (Custer 1982a), and Mitchell Farm (Custer 1981) in Delaware; Minguannan (Wilkins 1976) and Webb (Custer n.d.) in Pennsylvania; Indianhead (Mounier 1975), Raccoon Point (Kier and Calverly 1957), and Woodbury Annex (Cross 1941, 67–71) in New Jersey; and Hollingsworth Farm (Thomas 1982) and Conowingo (McNamara 1982a, 1983a) in Maryland. The only excavation data from microband base camps come from the Green Valley site complex (Custer et al. 1981) and no data are available for procurement sites. The presence of features such as house patterns, storage/refuse pits, and sheet middens helps us understand social complexity and community organization. None of the above-listed sites had unequivocal Late Woodland subsurface features or sheet middens. When features were found, they either dated from earlier time periods or their purported Late Woodland context was suspect. We discuss each of these instances below.

The Delaware Park site yielded numerous storage pits (Thomas 1981). Radiocarbon dates and diagnostic artifacts place these features in Late Archaic through Middle Woodland contexts. A limited number of features are also reported from the Crane Hook site (Swientochowski and Weslager 1942) but they apparently date from before Late Woodland times. Numerous features were excavated at the Clyde Farm site (Custer 1982a) and are dated to the Late Archaic through Middle Woodland periods by diagnostic artifacts and soils data. Pits occur at the Woodbury Annex site (Cross 1941, 67–71), with Middle Woodland affiliations suggested by associated ceramics. The Hollingsworth Farm site (Thomas 1982) is the last site with reported features and is the most enigmatic. Some features clearly are from the pre-Late Woodland context based on associated diagnostic ceramics. Also reported are other features of questionable origin. Pedological analysis indicates that the soil anomalies that characterize these "features" were caused by tree falls (Thomas 1982, 21–23). Thomas (1982, 21, 24) maintains that the large number of artifacts found in these soil anomalies indicates that the tree falls disturbed cultural features. Because these features contained Minguannan ceramics, they may be

Late Woodland pits. However, the statistical analysis of the features presented by Thomas is insufficient to support this contention.

Interpretations

The absence of pit features, house patterns, sheet middens, and burials is consistent with the settlement pattern data, which indicate the absence of sedentary villages or base camps. Even more interesting is the fact that Late Archaic through Middle Woodland occupations of the same macroband base camps do contain features including storage/refuse pits, hearths, and semisubterranean pit houses. Differential preservation and disturbance by plowing may account for the absence of some features. However, we suggest that not only were Late Woodland groups no more sedentary than preceding cultures, but they may have been less sedentary.

Consideration of subsistence patterns is difficult in light of the absence of preserved subsurface features. However, comparisons with earlier deposits from the same sites and with other areas provide some insights. The only analysis of flotation remains for Woodland sites in the study area is the work of Crabtree and Langendorfer (1981) with seed remains from the Delaware Park site. A wide variety of wild plant foods were identified, with amaranth and chenopodium important elements of the assemblages. A similar food base consisting of the intensive gathering of wild seed plants supplemented by hunting may be projected for Minguannan complex groups. The presence of extensive wild plant food remains at sites in the Lower Susquehanna (Ameringer 1975) and Upper Delaware (Moeller 1975) valleys, where groups were clearly using cultigens such as maize and squash during Late Woodland times, further supports this contention.

Building from the settlement pattern data and projections of the subsistence base, one can comment on social complexity, with amplification from the ethnohistoical record. Although a subsistence base of intensively gathered plant food seems to have supported ranked societies in southern areas of Delaware during Early and Middle Woodland times (Custer 1982b, 1983), the attendant environmental circumscription that caused these developments was not present in the Upper Delmarva Peninsula and Piedmont Uplands, where simple egalitarian organizations existed during Late Archaic through Middle Woodland times. Because Late Woodland societies were no more sedentary (and in fact even may have been less so) and probably had comparable subsistence bases, it is likely that there were simple egalitarian organizations during these times as well. Compound

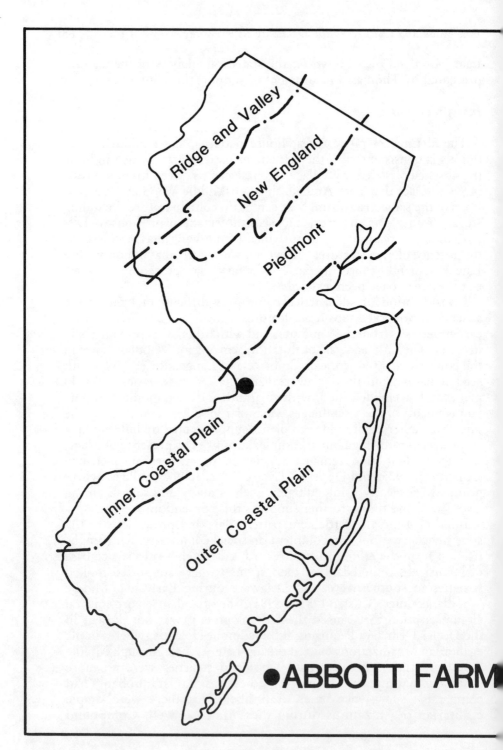

Fig. 6. Abbott Farm location.

bands were probably the largest social units and a nuclear family or set of families was probably the basic social unit of production and consumption. A detailed consideration of ethnohistoric data is beyond the scope of this paper; however, descriptions of social organization, the subsistence base, and settlement patterns derived from Late Woodland archaeological data are consistent with ethnohistoric analyses of local Lenape culture (Weslager 1953, 1961, 1968, 1972; Becker 1976a, 1980).

Coastal Plain/Piedmont Transition: The Abbott Farm and Vicinity

Background

The transition area from the Inner Coastal Plain to the Piedmont is a unique environmental region in the Delaware River Valley (fig. 6). It represents the upriver extent of the tidal influence and associated freshwater marshes. This location was historically the focus of seasonal runs of anadromous fish, a situation that probably lasted throughout the Late Woodland period. Floodplains are generally broad. Steep, ancient high terraces of the Delaware River characterize the transition to the undulating topography of the uplands. Sites of the Abbott Farm National Landmark near Trenton, New Jersey, crosscut each of these topographic zones. Data from the Abbott Farm form the basis of the following analysis, which is supplemented by site data from Delaware Valley areas north of Philadelphia to just north of Trenton. The majority of this area is similar to the Abbott Farm and its vicinity in that it possesses freshwater tidal streams, marshes, and associated terrestrial habitats.

The Abbott Farm and its complex of sites became prominent during the late nineteenth century when Dr. C. C. Abbott reported finding evidence of "paleolithic man" in the high-terrace gravels along the Delaware River (Abbott 1872). Work continued into the early twentieth century with the excavation of large areas of upland and lowland sites (Volk 1911; Skinner 1914, 1915; Spier 1918). The excavation of 1930s and 1940s by Dr. Dorothy Cross and the New Jersey Indian Site Survey at the Abbott Farm and throughout the Delaware Valley are probably the most well known (Cross 1941, 1956). Recent cultural resource management studies have continued to add to the substantial data base of this region (c.f. Pollak 1968, 1977; R. M. Stewart 1981b; Berger & Assoc. 1983a, b). The Abbott Farm and vicinity are currently the focus of a multistage investigation by Louis

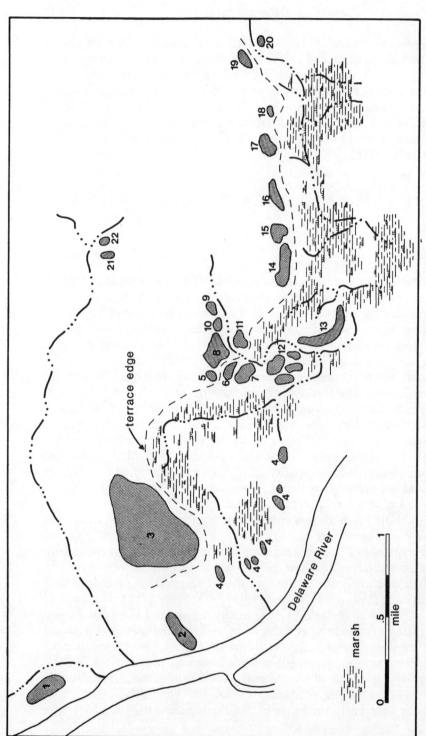

Fig. 7. Late Woodland sites of the Abbott Farm locality. 1: Industrial Terrace; 2: Riverview Cemetery; 3: Lalor's and Wright's Field; 4: Sturgeon Pond complex; 5: Watson House; 6: Excavation 14; 7: Roebling Park; 8: Excavations 2–4, 9, 10, 12; 9: Independence Mall; 10: Abbott's Brook; 11: Abbott's Lane; 12: Wetland Area sites; 13: Area B; 14: Lister; 15: Carney Rose; 16: White Horse; 17: Bordentown; 18: Site "G"; 19: Gropp's Lake; 20: Site "A"; 21: 28ME99; 22: Shady Brook.

Berger and Associates in advance of proposed highway construction. New fieldwork and reevaluations of past research have identified surface, plowzone, and buried Late Woodland deposits at multicomponent sites. In the case of buried deposits, Late Woodland materials are often mixed with diagnostic artifacts of the Middle Woodland period. Assignment of sites to the Late Woodland period is based upon pedological studies, stratigraphic sequences, and the occurrence of triangular projectiles and regionally known ceramic types. Figure 7 depicts relevant site distributions, which include surface, plowzone, and buried deposits.

Material Culture

The second humus at Excavation 14 (Cross 1956) and buried deposits at the Shady Brook (Stewart 1981b, 1984) and Gropp's Lake (Berger & Assoc. 1982a) sites provide the most reliable contexts for evaluating late Middle Woodland/Late Woodland assemblages.

Except for stylistic changes, the stone toolkit remains that of earlier Woodland times and reflects the functional diversity associated with exploiting a broad resource base. The use of argillite, presumably from the primary outcrops that begin five miles to the north of the Abbott Farm (Didier 1975), continued into the Late Woodland period but at a rate that was significantly lower than that of earlier times (Berger & Assoc. 1982b, 34). The degree to which argillite was used varies, depending on the type of site being examined. Cryptocrystalline lithics, as well as other materials, are more frequently represented. Staged bifacial reduction appears to have coexisted with a core and flake technology. A variety of bone and antler tools were also used during this time, including antler harpoons with drilled holes and antler tine projectiles (Cross 1956, 118–20, 196).

The use of copper for ornaments continued into the late Middle/Late Woodland from at least the Early Woodland period at the Abbott Farm. Researchers found rolled copper beads in the second humus at Excavation 14 (Cross 1956, 121), and a rolled copper tube also turned up in Pit 3 at Excavation 13, in association with a triangular projectile and Iroquoian-type ceramics (Cross 1956, 46; n.d., arrowhead type 1).

Ceramic assemblages for the period are diverse. Middle Woodland Abbott zone-decorated ceramics probably continued into Late Woodland times (Stewart 1982b, 24). Abbott fabric-impressed ceramics were recovered from the second and third humuses at Excavation 14, although they were three times more prevalent in the second humus (Cross 1956, 151). This type resembles both Riggins fabric:

impressed and overpeck incised (Cross 1956, 151), which one more typically associates with the Late Woodland period (Kraft and Mounier 1982, 160–62) and which was also found at Excavation 14. Other regionally known Late Woodland ceramics from the Abbott Farm include the Bowman's Brook, Indian Head, Owasco, and "proto-Iroquoian" types (Cross 1956, 134–64, 185). The shell-tempered "Andaste" ceramics noted by Cross (1956, 157–59, plate 41c) are the equivalent of the Schultz incised type associated with late-sixteenth-century Susquehannock Indians (Kinsey 1959, 95). Cross (1941, 107–17,167) also examined two sites in the Trenton vicinity that possessed shell-tempered, Iroquoian-styled ceramics. Fragments of a Schultz incised vessel were excavated at the Shady Brook site, on the edge of the Abbott Farm (Stewart 1981b, 78).

Settlement Patterns

The following analysis is based largely on data from the Abbott Farm complex of sites and immediately adjacent areas (Volk 1911; Skinner 1914, 1915; Spier 1918; Cross 1941, 103–7; 1956; Pollak 1968, 1975, 1977; Stewart 1981; Stanzeski 1981; Berger & Assoc. 1983a, b; File Reports, Office of Cultural and Environmental Services, N.J. Dept. of Environmental Protection; Dr. Jacob Gruber, personal communication). Additional Coastal Plain data come from Cross (1941, 95–103, 107–132) and Kraft and Mounier (1982, 158–73).

Late Woodland sites often occurred with the following types of stream environments:

—river and marsh associations;
—junction of river with streams of any order;
—associations of marsh with second- or higher-order streams;
—junctions of first- or higher-order streams;
—junctions of second- or higher-order streams with extinct or seasonal drainages;
—drainage headwaters, including springheads.

In all cases, site areas characterized by well-drained, low-relief topography correspond with the noted hydrologic associations. There have been no systematic surveys reflecting the environmental diversity of the region (Kraft and Mounier 1982, 1973). However, the above list indicates that a diverse range of habitats was the focus of Late Woodland settlement and subsistence. The spacing of sites appears to vary according to the juxtaposition of environmental features and associated resources. At the Abbott Farm, for example, sites on the edge of the high-terrace bluff are closely spaced and tend to blur into one another (see fig. 7). Along the slope of the high terrace bluff are numerous springs that feed a freshwater, nontidal marsh at the base of the bluff slope. Beyond this marsh are freshwater tidal marshes associated with tributary streams of the Delaware River. In part, high-terrace site distributions reflect the proximity of lowland marshes and their broad extent. They also reflect the nature of upland plant and animal resources that could have been exploited. Major upland resources, relatively widespread, would have consisted of the nuts and acorns of deciduous trees and mammals such as deer and squirrel. Furthermore, the numerous springs along the high-terrace slope would have ensured that the availability of fresh water would have been a factor limiting settlement choice.

The functional site types for the Late Woodland period discussed below refer specifically to the Abbott Farm and its adjacent areas (fig. 7). However, this typology applies to the broader region. Existing provenience data, incomplete analyses, and complex stratigraphy make it difficult to integrate many of Cross's 1956 upland site excavations into this typology. Site extent, depositional intensity, artifact and feature variety, and seasonality and duration of occupation are the criteria used for distinguishing site types. This approach is based upon site typologies previously employed by Gardner (1977, 1978, 1982) elsewhere in the Middle Atlantic region.

> *Macroband Camp:* An intensely reused or sedentary habitation site serving as a base for hunting, gathering processing resources, manufacturing, and social activities. Does not appear to have been inhabited on a strictly seasonal basis or for specialized activities. Artifact and feature variety higher than other site types. Features include caches, hearths, burials, and storage/refuse pits. Band segments probably coalesced here.

Excavations 2, 9, and 14, the Industrial Terrace site, Lalor's and Wright's fields, Watson House, and Roebling Park can be tentatively grouped into this type for the Late Woodland period (see fig. 7). Many large probable Late Woodland pit features are at the sites. At Excavation 14 (Cross 1956) and Roebling Park and Lalor's and Wright's fields (Volk 1911), the widely distributed pits are not segregated from any presumed "living" areas, which may show the intense degree of site utilization. Probable Late Woodland burials, known from Excavations 2, 9, and 14 (Cross 1956, 60–67, 196), are not segregated from other activity areas. Preliminary testing at the Industrial Terrace site revealed probably Late Woodland burials (late soils context and triangular point association), although not enough of this site has been exposed to evaluate any patterning. Even though other macroband base camps have yielded burials, so far we cannot firmly link them with the Late Woodland period.

Analysis of faunal materials from Middle/Late Woodland deposits at Excavation 14 points to some form of habitation during all seasons of the year (Parris 1980; Williams, Parris, and Albright 1981). However, no house patterns definitely match those of the Late Woodland period. Even so, Pollak (1968, 84) excavated portions of two curved postmold patterns on the high terrace near the Watson House that may be related. The depositional intensity and feature variety at the macroband camps imply prolonged occupations and the probable use of shelters of some type. We can infer the macroband characteristics of this site type from the degree of depositional intensity and ceramic design motif diversity usually found at other site types, especially since, diversity of design motifs is one marker of group fusion and fission.

With the exception of the Industrial Terrace site and Riverview Cemetery, macroband camps occur in highly productive marsh/floodplain/upland transition zones. (The Industrial Terrace site and Riverview Cemetery are associated with broad low-terrace settings of the Delaware River.)

> *Transient Camp:* Can involve sporadic or frequent reuse. There is evidence for structures, although the use probably involved an encampment. Overall artifact variety can be great but the quantity of any one artifact class and the depositional intensity in general is low. Lithic fabrication generally limited to tool maintenance and production of expedient tools. Feature variety generally limited to hearths and lithic reduction areas. Site structure a series of small activity areas centered on hearth features that appear to represent discrete episodes of site use.

The above definition in large part results from the extensively excavated Shady Brook and Gropps Lake sites (R. M. Stewart 1981b, 1984; Berger & Assoc. 1983c; chap. 6, 7) and tentatively includes Abbott's Brook, Lister, Carney Rose, and the Bordentown Waterworks. There, tools and ceramic artifacts are clustered around hearth areas. Diversity in ceramic designs within a hearth area is low but diversity can be very marked between areas, a pattern that holds for general artifact variety. Middle/Late Woodland hearths contain unidentified calcined bone and charred hickory nut fragments, which suggests a fall occupation. Grape has also been noted. Moderate to low depositional intensity and intrasite patterning may indicate exploitation by individual families or small male groups. For the region as a whole, transient camps come from the broadest range of physical settings and represent the characteristic settlement type associated with low-order stream environments.

> *Stations:* Relatively isolated, low density lithic scatters primarily composed of debitage indicating tool maintenance. Broken and/or wasted tools rarely found. Discrete scatters generally composed of less than fifty pieces of debitage. Features extremely rare and when present limited to small hearths (adapted from CRG/LBA 1983, 11–54).

Stations apparently resulted from the activities of individuals or small groups on hunting forays. Low-density lithic scatters and rare occurrences of broken or wasted tools would coincide with the maintenance of knives and other tools employed in the hunt and field dressing of game. The list of sites similar to this type is extensive and includes Gropp's Lake (Area A), White Horse, Independence Mall, Abbott's Lane, all Sturgeon Pond sites, and all wetlands area sites (see fig. 7). Stations occur in a wide variety of settings, and their locations are associated with many specific environmental features.

So far there are no specialized Late Woodland procurement sites at the Abbott Farm. The procurement of cobble lithics from local secondary sources easily could have been embedded within other settlement and subsistence activities. Upriver from the Abbott Farm, specialized procurement sites may exist in association with argillite outcrops. Downriver and/or Coastal Plain sites may reveal an emphasis on shell fishing or salt marsh exploitation since the remains of salt marsh turtle and swan were found in the second humus at Excavation 14 (Williams, Parris, and Albright 1981). Specialized procurement sites may or may not involve habitation. Although a variety of resources may be exploited here, artifacts and floral and faunal remains show a quantitative focus on a single resource. Otherwise, this site type could resemble a transient camp.

Kraft and Mounier (1982, 42) find that systemic relationships among the Late Woodland sites in the central and lower Delaware Valley have been consistently overlooked. Nevertheless three models can integrate the available data. The first one (fig. 8), essentially a central-based wandering system, proposes that some form of group was in year-round residence at a macroband camp. The floral and faunal data from the second humus at Excavation 14 support this contention (Williams, Parris, and Albright 1981). All other types would have occurred as forays from the base camp and may have been conducted on a seasonal or "as-needed" basis. Ethnohistoric data suggest that Delaware Indian populations were especially mobile in the summer (Goddard 1978a), and floral remains from Gropp's Lake

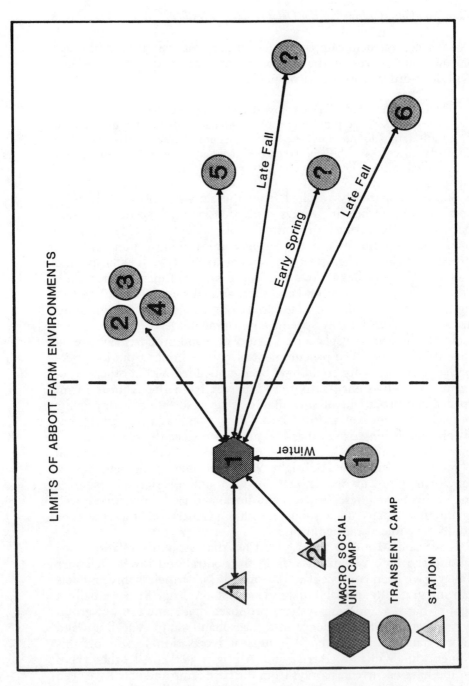

Fig. 8. Late Woodland settlement Pattern no. 1. *Macro-social unit camps:* 1. Excavation 14 or Lalor's and Wright's fields. *Stations:* 1. Abbott's Lane; 2. Area B. *Transient Camp:* 1. Lister; 2. Argillite procurement site; 3.

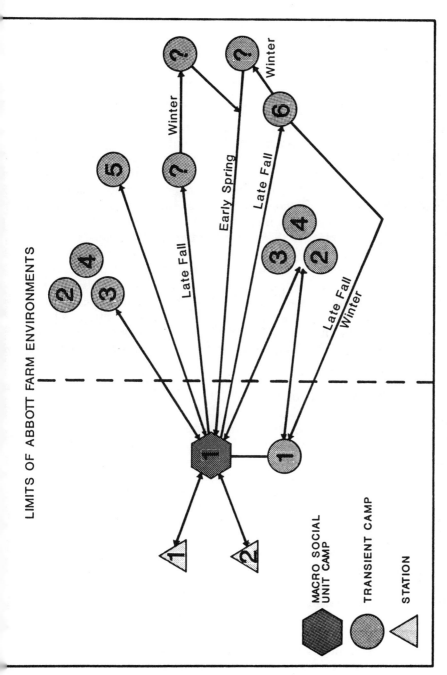

Fig. 9. Late Woodland settlement pattern no. 2. *Macro-social unit base camp*: 1. Excavation 14 or Lalor's and Wright's fields. *Stations*: 1. Abbott's Lane; 2. Area B. *Transient Camps*: 1. Lister; 2. Argillite procurement site; 3. Shellfishing site; 4. Mica procurement site; 5. Gropp's Lake; 6. Shady Brook.

(Area G) and Shady Brook indicate that fall forays were probably common. Over a period of years, the location of the base camp may have shifted in response to diminishing resources, population pressures, or political realignment. Thus settlement would have alternated between a number of macroband camps.

The second alternative (fig. 9) proposes that macroband base camps disbanded on a seasonal basis with individual families or family groupings moving to other camps where the functions performed and the duration of occupation would have depended on the resource base. Various degrees of group fusion and fission would have continued to have occurred away from the base camp. This model appears to be the most consistent with ethnohistoric data (A. Wallace 1949; P. Wallace 1961, 76; Hunter 1959, 15; Goddard 1978a, 216). However, the nature of the Late Woodland site types indicates that postbase camp residence may have involved more than hunting and the single family exploitative unit suggested by ethnohistoric data. Faunal remains representing all seasons at Excavation 14 could have resulted from small group forays back to the area after the base camp had been abandoned by the macroband. Seasonal or annual regrouping may have involved locations that were the focus of macroband camps at other times of the year.

The third possibility (fig. 10) assumes that the location of the macroband camp shifted during the year between interior marsh settings and broad, well-drained Delaware River terrace environments. Cavallo (1982) suggests that shifting to broad floodplain areas during the later portions of seasonal anadramous fish runs would have increased fish yields.

Late Woodland sites at the Abbott Farm and in the surrounding region are multicomponent and generally include Late Archaic through Middle Woodland deposits. Although actual site locations remained relatively consistent from the Late Archaic through the Late Woodland, the function of some individual sites changed. In the Abbott Farm wetlands a functional shift appears to have occurred during the Middle Woodland period (R. M. Stewart 1982a, 26–27) and lasted into Late Woodland times (R. M. Stewart 1982b). The location of major Late Archaic/Early Woodland occupations (fig. 7, area B) is not the location of major Middle and Late Woodland deposits (fig. 7, Excavation 14 and Roebling Park). By Late Woodland times, Area B is merely the scene of sporadic small-group forays. The change in site locations during the Middle Woodland period may have resulted from upstream shifts in tidal influence affecting anadromous fish runs in conjunction with an intensification of subsistence production (R. M. Stewart 1982a, 26–27; R. M. Stewart 1982c).

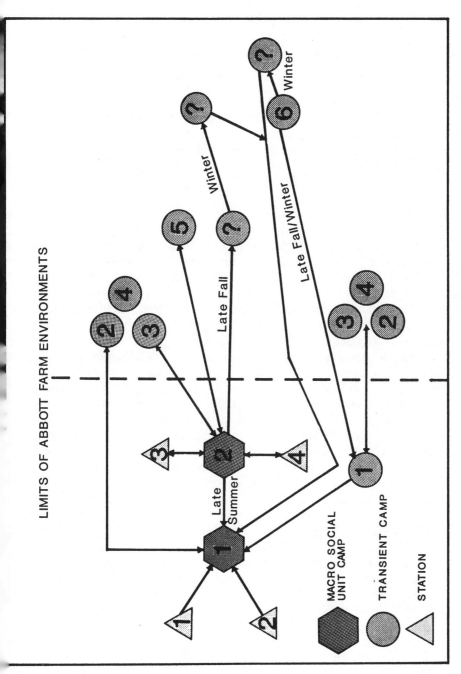

Fig. 10. Late Woodland settlement pattern no. 3. *Macro-social Unit Camps:* 1. Excavation 14; 2. Lalor's and Wright's fields. *Stations:* 1. Abbott's Lane; 2. Area B; 3. Independence Mall; 4. Sturgeon Pond. *Transient Camps:* 1. Lister; 2. Argillite procurement site; 3. Shellfishing site; 4. Mica procurement site; 5. Gropp's Lake; 6. Shady Brook.

During Late Woodland times, gradual sea level rise would have continued to alter tidal patterns within the Abbott Farm wetlands. More upstream portions of wetlands streams may have become the best places to exploit anadramous fish runs. Lalor's and Wright's fields and the Riverview Cemetery would have been locations from which altered marsh, traditional floodplain, and upland habitants could have been simultaneously exploited. Similar patterns have been documented in the Maryland (R. M. Stewart and Gardner 1978) and Delaware (Williams and Thomas 1982, 124) Coastal Plains.

Subsistence

A variety of floral and faunal remains have been identified from probable late Middle Woodland/Late Woodland contexts at the Abbott Farm. Excavations at Shady Brook (R. M. Stewart, 1981b; R. M. Stewart n.d.) produced charred grape and hickory nuts, together with fragments of calcined bone primarily from small mammals. Charred hickory nuts are also in the burial deposits of Area G at Gropp's Lake (Berger & Assoc. 1982a, chap. 6). Volk recovered deer, unidentified bird, and chestnut and hickory nuts from pits at Lalor's Field, which also contained only triangular projectile points (Volk 1911, 1890 to 1893 excavations—Trench 6; Trench 9, Pit 2; Trench 10, Pit 1; 1895 excavations—Trench 2, Pit 3). Salt marsh turtle, sturgeon, otter, swan, goose, duck, turkey, deer, bear, elk, beaver, raccoon, and wolf or dog have been identified from the second humus at Excavation 14 (Williams, Parris, and Albright 1981). Volk's (1911, 67–68) nearby excavations at Roebling Park produced similar remains from pits in addition to catfish, hickory, and walnut. However, the Late Woodland affiliation of these specimens is not clear. White-tailed deer is the most important meat source found at Excavation 14, followed by elk, sturgeon, bear, beaver, turkey, and raccoon (Williams, Parris, and Albright 1981). Differential preservation and difficulties in estimating minimum numbers of individuals, especially in the case of sturgeon, may have biased these representations (Williams, Parris, and Albright 1981; Cavallo 1982). Floral and faunal remains indicate a broad Middle/Late Woodland resource base that is also implied by the diverse settings in which sites occur. However, it would be premature to suggest the degree to which hunting, gathering, and fishing contributed to subsistence.

Seventeenth-century accounts indicate that maize to some degree was a part of Delaware Indian subsistence (Newcomb 1956, 13; Goddard 1978a, 217). Whether or not the use of cultigens played a role in traditional lifeways or was the result of European acculturation is still in doubt. No physical remains of prehistoric cultigens are known from the Inner Coastal Plain of New Jersey (Kraft and Mounier 1982, 162). Excavations at the Turkey Swamp site, located in the Outer Coastal Plain approximately twenty miles from the Abbott Farm, produced charred maize kernels from Late Woodland levels (John Cavallo, personal communication). Charred maize has been radiocarbon-dated to circa A.D. 1300 at the Williamson site, approximately thirty miles from the Abbott Farm (see below).

The absence of cultigens at the Abbott Farm is especially surprising, given the massive excavations by Volk (1911) and Cross (1956), which produced other plant food remains. Lithic artifacts that could have been used in the cultivation and processing of cultigens (digging sticks, hoes, mortars, pestles, mullers, grinding platforms) occur with equal frequency in Late Woodland assemblages, as they do in earlier cultural deposits. Some of these items may have been fashioned from wood (c.f. Goddard 1978a, 217), with poor preservation accounting for their absence and the lack of cultigens. By themselves, the use of certain artifact types as indirect evidence for the use of cultivated plants will always be ambiguous. Alexander (1969, 124) notes that there are ethnographic examples of societies using digging sticks, milling equipment, and storage pits in the absence of domesticated plants.

Maize-based agriculture could have been practiced merely as a seasonal sideline. Cultigens may simply have been items added to an undoubtedly extensive list of other plant foods that had been exploited throughout Late Archaic to Middle Woodland times. Domesticated plants could have been raised in a "garden" situation at almost any type of site. In this scenario, the small amount of food produced would have served a minor role in subsistence pursuits. The broad lower terraces of the Delaware River are ideal locations for prehistoric horticulture on a larger scale—far better in fact than the circumscribed floodplain zones of Watson's and Crosswick's creeks. The Industrial Terrace Site and Lalor's and Wright's fields could represent agricultural villages in this case.

Social Organization

During the Middle Woodland at the Abbott Farm, there is evidence for the existence of ranked societies based on the link between social organization and subsistence intensification. Intensification of the mode of production has been theoretically and ethnographically linked with modified social organization and ranking (Price 1981, 1982). Intensification is thought to have involved an increased focus on anadromous fish in combination with the simultaneous exploitation of upland resources. The proposed intensification is archaeologically visible by settlement shifts, which would have facilitated simultaneous lowland/upland exploitation; and increase in the production of argillite bifaces and caches, tools presumably important in fish processing; and an increase in storage/refuse features and ceramic storage vessels (R. M. Stewart 1982a, 26–27; 1982c; Cavallo 1982). At the Abbott Farm, ranking appears to have been at the level of a "Big Man" society (R. M. Stewart 1982c; Sahlins 1958). This type of social organization may have also characterized populations for an undetermined portion of the Late Woodland period. Vague indications of status burials from the second humus at Excavation 14 and from pits at Excavations 2 and 9 lend some support to this hypothesis. Skeleton 60 at Excavation 14 is a flexed burial that contained four triangular projectiles beneath the skull and sixty-six scrapers piled at the feet with an Iroquoian-styled pipe (Cross 1956, 66, 196). A single-flexed burial from Excavation 2 contained triangular projectiles and deer antlers that were placed above the skull (Cross 1956, 63, Skeleton 11—Pit 13). Single burials with deer antlers are also known from Excavation 9 and the second humus of Excavation 14 (Cross 1956, 65, 67; Skeleton 40—Pit 59; Skeleton 82—Pit 84).

This type of social organization may have been eventually disrupted by an emphasis on cultigens in Late Woodland subsistence. Primitive agriculture would not have required the organization and intensive labor associated with the earlier intensive exploitation of anadromous fish. Although we do not know the exact nature of the new social order, it does not appear to have involved the low-level ranking of Middle Woodland times. Early historic records portray the Delaware Indians of the study area as a collection of autonomous bands with no overriding organizational framework (Hunter 1959, 15).

The Middle Delaware Valley

As used here, the Middle Delaware Valley (see fig. 1) conforms to Kinsey's definition as the region between the Water Gap and Trenton (Kinsey 1975, 8). Trenton is near the Fall Line, the escarpment that marks the edge of the continent. In the Middle Delaware Valley the river cuts through three physiographic provinces: Ridge and Valley, Highlands, and the Piedmont. The southern half of the valley is within the Piedmont. In the Piedmont, the valley becomes wider, with fewer cliffs bordering the river.

The following discussion suggests that during the Late Woodland

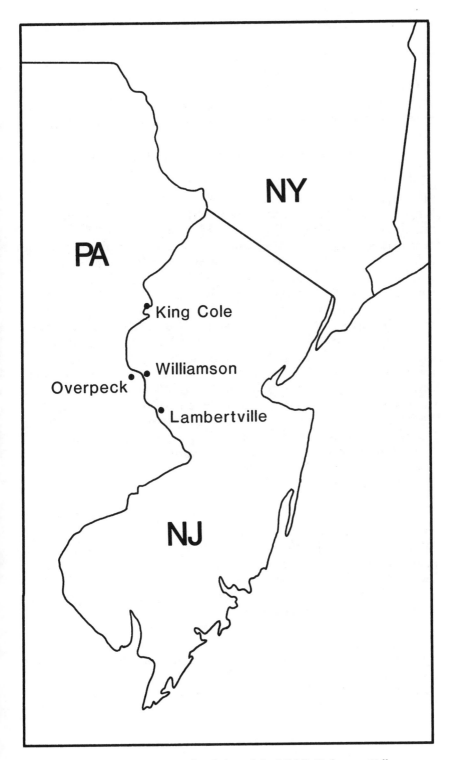

Fig. 11. Selected Late Woodland sites of the Middle Delaware Valley.

period in the Middle Delaware Valley settlement systems had certain characteristics: (a) as in earlier times, the systems remained focused on floodplains with scattered, semisedentary base camps in evidence; (b) they used satellite camps such as rockshelter; (c) they experienced an inland shift of base camps late in the period; (d) primarily depended on hunting and gathering economy on which agriculture had a minimal and late impact; and (e) were comprised of egalitarian bands.

Few sites have been thoroughly investigated in the Middle Delaware Valley. The Overpeck (Forks of the Delaware Chapter 1980) and Byram (Kinsey 1975) sites are probably the most well known. Other notable sites include Williamson (Hummer 1978), Lambertville (Struthers and Roberts 1983), Sandt's Eddy (Fehr et al. 1971), King Cole (F. Dayton Staats, personal communication), Walter's Nursery (Landis 1982), Washington's Crossing (Dr. A. Ranere, personal communication), and the Upper and Lower Black Eddy sites (Mercer 1897). Buried and radiocarbon-dated Late Woodland components are rare in this listing. A date of A.D. 860–80 (1070, MASCA corrected) was obtained from a feature at Lambertville but does not apply to the major Late Woodland component represented by Sackett corded ceramics (Struthers and Roberts 1983). Overpeck produced two Late Woodland occupation levels but no radiocarbon dates. Williamson has a single Late Woodland date of A.D. 1290 ±80 (DIC–875, MASCA corrected to A.D. 1220 to 1360). Portions of intact living floors have also been excavated at Williamson. In figure 11 are selected Late Woodland sites in the Middle Delaware Valley.

Settlement Patterns

As yet, no systematic survey has been undertaken along the bluffs, in the hollows, and on the uplands that form the borders of the Valley. There is evidence of Late Woodland use of rockshelters as probable microband procurement camps in adjacent areas. The Unami Creek Rockshelter (Strohmeier 1980) and the Thousand Acre Rockshelter (William Strohmeier, personal communication), both located in the Piedmont of Montgomery and Bucks counties, Pennsylvania, yielded Owasco (Sackett), Owascoid, and Overpeck paddle-edge stamped ceramics, among others, together with numerous Late Woodland triangular projectile points. These ceramics are also common on floodplain sites. According to Strohmeier (personal communication), the presence of Late Woodland materials on inland Piedmont sites in Bucks and Mongomery counties is minimal. Scattered triangular points turn up on open sites but are never prominent; moreover, the sites always possess earlier components. With the exception of one site on Unami Creek, Late Woodland ceramics are rare on open sites but are a standard feature of rockshelter assemblages.

The Dark Moon site (Patricia Hartzell 1982, personal communication), located on the poorly drained floodplain of a small tributary of Beaver Creek, is an inland site in the Ridge and Valley section of southern Sussex County, New Jersey. Although outside the study area, it is significant as an inland site with probable evidence for sedentism in the form of house patterns. To date, five oval or round patterns of postmolds have been located. The house patterns range in size from 15 by 15 feet to 22 by 35 feet. In addition, twenty-one small pits have been excavated, none of which is comparable to the deep silolike features found on Late Woodland sites several miles to the west in the Upper Delaware Valley. Twenty of the 21 pits contain deer bones. Ceramics are numerous but vessels are small in size. The thousands of triangular points found often have mends of broken fragments.

Although excavations at Dark Moon are incomplete, tenative interpretations can be offered. Based on the house patterns, the site would appear to be a sedentary camp with possible year-round habitation. The absence of large below-ground storage features and storage-size ceramic vessels, its location adjacent to black flint outcrops, and the on-site occurrence of significant amounts of black flint chipping debris all suggest a specialized procurement site that may also have served as a microband base camp.

Archaeological investigations of Middle Delaware River floodplain sites (Forks of the Delaware Chapter 1980; Hummer 1978; Struthers and Roberts 1983; F. Dayton Staats, personal communication) support the hypothesis that Late Woodland peoples lived in small extended family bands in dispersed semipermanent to permanent camps (Becker 1976, 1980a; Kraft 1975a, 1978). Excavations and surface surveys in the Riegel-Compton-Williamson site complex south of Milford, Hunterdon County, New Jersey, have revealed many dispersed floodplain sites characterized by small scatters of Late Woodland ceramics and projectiles. The Williamson site has been under excavation by Hummer for more than ten years.

Williamson is a multicomponent site on a river levee 8.5 meters above the normal summer flow of the river. The levee is elevated above the inner floodplain and has better drainage. The bulk of the inner floodplain is a poorly drained floodchute that extends south through the site to Harihkake Creek, a distance of about three kilometers. During seasons of heavy rainfall a substantial pond forms in the floodchute east of the site. This feature increases the resource potentials of an already-rich floodplain environment (Godfrey 1980, 260, 302). Because of the ecology of temperate zone floodplaines, it is possible that some type of aboriginal subsistence activity could have been pursued year-round.

Investigations in the Riegel-Compton-Williamson complex suggest that a focused floodplain adaptation began during the Late Archaic/Early Woodland and persisted through Late Woodland times (Hummer n.d.). Postulated warm/dry climate extremes during the Late Archaic period (Carbone 1982, 45) may have in part helped to initiate this settlement/subsistence system (Custer 1978, 1980, 1982b; Custer and Wallace 1982, 147). The wide distribution of attractive floodplain environments in the study area corresponds with the dispersed Late Woodland site distributions noted above. The hunter-gatherer model proposed by Jochim (1976) and applied by Stevenson (1982) to settlement pattern analysis in Centre County, Pennsylvania, may fit the Middle Delaware Valley data with some modifications. The general continuum affirmed by both Stewart (1982a) and Custer (1982b) for the period circa 1000 B.C. to A.D. 1000 in lower segments of the valley may also be valid for the present study area and would include the Late Woodland period. No major Late Woodland adaptive shift to horticulture/agriculture (Thomas 1981, III-4) is evident in the Middle Delaware Valley. The major adaptive shift occurred early in the Woodland period with the initial concerted move to the floodplains. A minor shift may have occurred in late Late Woodland times with the relocation of some base camp sites to inland zones.

At Overpeck, Lambertville, Byram, Williamson, and Site B at the Riegel-Compton-Williamson complex, Late Woodland occupations overlap with earlier Woodland components. Settlement systems appear to have been conservative through time. Sites like Williamson persisted as relatively small, semisedentary base camps. The variety of features at Williamson is additional evidence for sedentism. Features include different types of small hearths, pits, lithic workshop areas, ceramic vessels of all sizes including some of storage size, and a fairly large and varied tool assemblage. Pit size is small to moderate in proportion (less than 1 meter deep). Missing at Williamson and apparently at Overpeck are the large storage pits found on Late Woodland sites in the Upper Delaware Valley (Kraft 1978, 27). Inferred sedentism for the Middle Delaware Valley does not seem to be connected with reliance on cultivated plants but on a highly successful hunting and gathering system predating the occurrence of cultigens.

Subsistence

At Williamson there is evidence for agriculture circa A.D. 1220 to 1360 in the form of radiocarbon-dated (DIC-875, MASCA-corrected) fragments of charred maize recovered from refuse pits. R. M. Stewart (see above) has addressed the puzzling absence of ground stone tools for use in plant processing on area sites. A chipped stone hoe was excavated from the plowzone at Williamson but grubbing tools need not be associated solely with the use of domesticated plants (Alexander 1969, 124). An emphasis on hunting is suggested by high projectile density ratios at Williamson (1:50 square feet), Lambertville (1:81 square feet) (Struthers and Roberts 1983, 76), and Washington's Crossing (Dr. A. Ranere, personal communication).

Charred nutshells, especially hickory and walnut, are abundant in the excavated pits at Williamson. Further evidence for a basic hunting and gathering economy is the presence of netsinkers (29), calcined bone from small animals and fowl, charred seeds of several varieties (as yet unidentified), and quantities of deer bone. Bones from all parts of the animal are represented, indicating that entire carcasses were brought back to the site for butchering.

The Overpeck site (Forks of the Delaware Chapter 1980) yielded a large number of triangular projectiles (156), although density ratios are not available. Also included in assemblages are ninety-five netsinkers, several pestles, grinding platforms, and a good association of teshoas and deer bone. The abundant nuts include walnut, butternut, hickory, acorn, and chestnut. Maize cobs, kernals, and charred beans are also reported but cannot be confidently assigned to either of the two Late Woodland components at the site.

Maize-based horticulture/agriculture appears to have been a relatively late introduction and had little impact on traditional settlement/subsistence systems. The existing environment was productive and population levels may not have been critical, which would have required alterations or shifts in traditional strategies. The minor role played by maize is supported by ethnographic sources (Newcomb 1956, 13) and by the situation in the Upper Delaware Valley. Out of 411 refuse pits excavated at Bell-Post-Philhower, only four contained maize, and they are attributed to the Minisink/Historic time frame (Kraft 1978, 44–45). Kraft (1978, 28) states that

> . . . it cannot now be determined how great a dependence the peoples of the Minisink area had on horticultural foods. It is obvious that the local environment provided a substantial economic input.

Kinsey (1975) presents data that suggest an almost consistent increase in probable equipment at the expense of hunting gear during later years of the Late Woodland (Owasco through Tribal). The data are not unassailable, however, even by Kinsey's admission (1975, 25, 26). The impact on agriculture appears to have been late for the Upper Delaware Valley, and for the Middle Delaware Valley it may have been even later.

Social Complexity

Data from Williamson and Overpeck tentatively indicate that Late Woodland society was neither well integrated nor rigidly structured. Intrasite patterning at Williamson suggests that a single component may be represented by site deposits. As mentioned earlier, Williamson is positioned on a levee with topography sloping sharply away from the site to the west and to the south. On the southern slope is a cluster of nearly two dozen refuse pits. This uncluttered arrangement of features and debris is a classic example of discard patterning at relatively sedentary sites and conforms to ethnographically derived data on Delaware Indian discard behavior (Murray 1980). Such orderly patterning might not be expected if the site had been repeatedly occupied by groups unfamiliar with an earlier layout.

The cluster of refuse pits contained the remains of at least a dozen different vessels. Several types occur in these pits: Castle Creek puctuate, East River cord-marked, Owascoid corded horizontal, and two untyped vessels. Adjacent pits in the cluster contained Owascoid corded collar and Overpeck paddle-edge stamped, plus untyped vessels. Cultural affinities seem to be to the north and south, although Upper Delaware Valley ceramic types are in the minority at Williamson (see table 4).

Using data from New York State, Whallon (1968) developed a model in which ceramic homgeneity was an indirect result of increasing population pressure, an increasing dependence on agricultural products, and the development of complex social organizations among Iroquoian groups. Hypothesized population and political pressures led to changes in subsistence production and intergroup relations. More rigidly structured descent groups and titled men who formed a local governing council were the social response to alterations in the biosocial environment. A pattern of increasing stylistic homogeneity of ceramics through time corresponds with the other changes noted. Ceramic homogeneity was interpreted as a result of

Table 4
Mixture of Middle-Late Woodland Ceramics from the Middle Delaware Valley*

	WILLIAMSON	OVERPECK
COLLARED VESSELS		
Owasco Corded Collar	X	X
Owascoid Corded Collar	X	X
Overpeck Paddle Edge Stamped	X	X
Castle Creek Punctuate	X	X
Unidentified (Untyped)	X	X
COLLARLESS VESSELS		
Owascoid Corded Horizontal	X	X
Overpeck Incised	X (?)	X
East River Cord Marked	X (?)	—
Unidentified (Untyped)	X	X

*The chart does not reveal the fact that at Williamson unidentified ceramics account for 57 percent of the total. Similar data is not available for Overpeck. Disappointingly the ceramic yield at Lambertville was very small. Collarless vessels at Williamson outnumber collared vessels (29 out of 45, or 64 percent), and unidentified collarless (22) account for 49 percent of the total number of vessels for the site.

decreased contact between villages because of warfare and more strongly integrated groups within villages.

If Whallon's model is valid, then several facts argue against the existence of rigidly structured society and for egalitarian organization in the Middle Delaware Valley: (a) the lack of homogeneity in pottery; (b) the lack of indicators of warfare such as stockaded villages or burials revealing violent death (Kinsey 1972; Kraft 1974, 1975, 1978); (c) no indications of agricultural intensification; and (d) the absence of status types of burials prior to European contact (Forks of Delaware 1980; Kraft 1975a, 1976a, 1978).

Summary and Conclusions

Late Woodland cultures of the Piedmont Uplands and the Upper Delmarva Peninsula show very little change from the preceding Middle Woodland period. Continuity of adaptations is best illustrated by settlement pattern data; however, site-specific research also shows basic similarities in the use of storage, artifact distributions, and community patterns.

The Fall Line area of the Upper Delmarva Peninsula, with its extensive marshes and highly productive habitats, provided a rich subsistence base where agriculture was not a viable alternative to intensive gathering supplemented by hunting. Also, Late Archaic through Middle Woodland data suggest that the extent of productive zones was so great that environmental circumscription was never a factor selecting for the development of complex societies (Custer 1982b). Similar arguments can be advanced for the Piedmont Uplands (Custer and Wallace 1982). The limited extent of well-drained floodplain settings in the Piedmont Upland would have limited the feasibility of agriculture.

At the Coastal Plain/Piedmont transition encompassed by the Abbott Farm, the eary Late Woodland also shows little change from the Middle Woodland period. Continuity is seen in material culture; settlement patterns, including sedentary base camps; an economic adaptation involving the intensive exploitation of anadramous fish in conjunction with hunting and gathering; and a "Big Man" type of ranked society. This adaptation was facillitated by a unique environmental setting combining circumscribed tidal marshes, broad well-drained floodplains; extensive primary and secondary lithic sources, and a diverse upland.

The possible reliance on cultigens circa A.D. 1300 represents a complex interaction of environmental and social variables (R. M. Stewart 1982b). The exploitation of productive marsh/floodplain/upland zones could have supported a large and growing aboriginal population. As increasing population began to stress the resource base, a number of adaptive options could have been selected. Populations could have splintered, new technologies could have been adopted or devised, the resource base could have been expanded to include new items, or traditional subsistence activities could have been intensified. During the Middle and early Late Woodland periods, the decision seems to have been to intensify subsistence production by simultaneously exploiting marsh, floodplain, and upland zones and by increasing and reorganizing labor. By A.D. 1300, recurring population pressure and/or the failure of the existing ranked society to manage labor and access to resources led to another social choice. A new technology of agriculture was adopted, the resource base was redefined to include cultigens such as maize, traditional subsistence pursuits took on a more supportive role, and populations splintered into discrete agricultural communities. This would have created a shift in the mode of production (Price 1981, 1982). Although we do not know the nature of post-1300 A.D. social organization, some form of the traditional social network would have had to be maintained or modified in order to reduce the subsistence risks associated with agriculture (Braun and Plog 1982, 508).

Above the transition to the Piedmont and below the Delaware Water Gap, scattered semisedentary macroband camps are in evidence, the use of storage is limited, and the adoption of agriculture circa A.D. 1300 had little impact on traditional lifeways. Social organization appears to have been egalitarian and at the level of the band. The earlier Woodland settlement system that was focused on the floodplain generally remained intact through the Late Woodland period. Base camps are small and dispersed, and the solution to any biosocial disruptions of the traditional adaptive system seems to have been the splintering of populations. The hypothesized inland movement of some base camp locations during the Late Woodland period may be one such "splintering." Minor climatic oscillations affecting the resource base (Carbone 1982, 46–47) and slowly growing populations may have constituted some of the disruptive threats to traditional lifeways. Ceramics indicate that resident groups participated in broad regional trends while maintaining what appears to have been a distinctive settlement and subsistence system.

The preceding discussions have addressed significant cultural variability in an area where Late Woodland peoples apparently shared a common language and material culture. To imply, as do some ethnohistorical records, that these groups also shared a common settlement/subsistence system and social organization is to deny the human race's basic relationship with its environment and accept a normative view of culture.

4
Cultural Diversity in the Lower Delaware River Valley, 1550–1750: An Ethnohistorical Perspective

MARSHALL J. BECKER

As the other papers that deal with the Delaware River Valley in this volume suggest, this region was occupied by a multiplicity of cultures whose adaptive strategies changed significantly in relatively short periods of time in response to environmental changes as well as to European Contact. The varied effects of European Contact within the Delaware River Valley were met by a wide range of strategies that may mirror patterns of cultural flexibility existing long before contact for local native American societies. Intensive archaeological research in the Delaware River Valley has identified localized patterns of archaeological complexes that strongly suggest the existence of cultural traditions within specific areas. This research also provides clear indications of change through time and of expansion and contraction of territories within which these people appear to have operated. These data from the Late Woodland Period (see R. M. Stewart, Hummer, and Custer, this volume) are vital in supporting the ethnohistoric information for similarities in cultural activity during the historic period.

As cultural consistencies within an area are being defined archaeologically and as the area is distinguished from other areas around it, archaeologists emphasize the reconstruction of the lifeways of the people who lived within the region. However, the nature of archaeological data is such that the limitations of information at one site are balanced with information derived from others that are considered to be related both in time and space (see Turner, this volume). One of the

---

The general research for this paper was initiated by a program funded by the American Philosophical Society (Phillips Fund) and continued through a grant from the National Endowment for the Humanities. The specific concerns of this paper were researched through a 1983 grant from the American Philosophical Society. Thanks are also due to Jay Custer, William A. Hunter, and Ronald A. Thomas for their many useful suggestions on an earlier draft.

values of ethnohistory is that it can provide clues to cultural behaviors that are difficult to recognize in the archaeological record. Also, ethnohistory often makes it clear that different clusters of individuals sharing the same culture may not operate their cultures in the same ways. Simply put, no two archaeological sites are identical. Often we believe that these differences may be due to temporal separation or environmental (ecological) adjustments to circumstances in the immediate neighborhood. Less often do archaeologists consider the possibility that the variations seen are the result of normative differences that can appear within a culture as the result of different kin groups and/or residential groups interpreting their supposedly similar culture in different ways. These cognitive differences may become more evident when we examine the range of variation among the various populations of a culture. Let us, then, examine historic data from the Middle and Lower Delaware River Valley, an area that until recently was considered inhabited by a single culture—the Lenape. This Lenape culture would be expected to produce archaeological sites roughly comparable from place to place within their realm. Yet it would also include the kinds of variation expected to occur within any constellation of related sites.

Note that until recently (Goddard 1978a) all of the Delaware Indians generally were lumped together as a single "people," although the Munsee and Lenape appear to represent two distinct cultures. Similarly, the extensive archaeological work conducted in the Upper Delaware Valley (e.g., Kraft 1978) was believed to provide data that would be shared throughout the Delaware Valley. The ethnohistoric evidence indicates that a considerable difference existed between these two provinces during the historic period. Archaeological information now available confirms earlier inferences on the cultural distinctions and one would expect that this will be further verified through future excavations.

It is important to note that even within the more restricted geographical area of the Lower Delaware Valley, the behaviors of specific bands identified within the Lenape tradition demonstrate a range of responses to changing circumstances that reflects incredible flexibility. Even within bands different strategies were being used to respond to the same stimulus. Since this is the case with the Lenape, a foraging people, we may be able to use this understanding of their variations in cultural responses to understand how various other peoples, such as the Late Woodland Monongahela or the contemporary people of Shenandoah Valley, met the challenges of their situation. What we usually see is the overall pattern, or what I have called the "mainstream" of cultural process. The data in this paper are intended to call attention to the range of variations in cultural responses within a community who shared, to one degree or another, the same culture, but who implemented their perceptions of the cultural rules in different ways.

Lenape and Jersey Indians

Within the area believed to be the territory used by the Lenape around 1600 A.D. a significant feature may be the Delaware River as a boundary between two clusters of native bands. Despite the belief by some historians that the peoples on both sides of the river constituted a single society sharing the same culture, there is only minimal evidence for the interaction of individuals across this modern and presumably ancient boundary. Aside from considerable interaction around the early 1630s (possibly circa 1610–50) when the Susquehannock (Minquas) asserted hegemony over the land between the Schuylkill and Christina creek drainages, little else can be documented. During this period of disruption, Lenape from southeastern Pennsylvania appear to have taken refuge on the east side of the river, although many may have moved either north or south out of harm's way.

The Susquehannock only had interest in this region as an overland route for getting furs to European markets, and later in maintaining good relations with colonial allies. Raids against the Lenape, to secure maize or other tangibles, appear to have been of secondary interest. We know almost nothing of interaction patterns between Lenape and their Jersey kin, but a few accounts suggest limited and friendly intercourse. The historic record, particularly clear after 1700 (see Becker, n.d. b), shows that the Jersey Indians had patterns of land sales and movement entirely different from those of the Lenape in Pennsylvania.

To a great extent the ecological differences between the Lenape area in Pennsylvania and the flat surface of southern New Jersey may have influenced the development of divergent cultural traditions. The difference between William Penn's methodical purchase of all Lenape lands in his Crown grant and the later (post-1700) system of securing title to land in the Jerseys (see Becker, n.d.b) may have served to emphasize ecological differences. The Jersey land sale pattern, being piecemeal and sporadic, may have encouraged the native residents (who are never noted by the term *Lenape* anywhere in the literature) to continue their traditional ways long after the Pennsylvania Lenape had left their ancestral area. However, those Jersey Indians who did leave tended to cluster together with their immediate kin and formed groups that largely maintained residence areas distinct from their more distant Pennsylvania relatives.

While the Lenape of Pennsylvania tended to move due west with some elements veering off to follow the Susquehannah River north, the Jersey people tended to move northwest. First the Jersey groups took up residence near the Forks of Delaware and then moved north and northwest into New York and eventually many moved to Canada. Much of the Jersey groups' post-1750 activity to some degree was indirectly in concert with that of the Munsee of northern New Jersey, although they were linguistically close to the Lenape of Pennsylvania. These differences, in the main, lead me to suspect that Goddard's (1978b) identification of two dialects of Lenape may correlate with these two populations. Thus the Late Historic terms used by various Lenape to refer to their own "ancestors" may reflect these Early Historic period distinctions. The term *Unami*, generally used after 1760 and believed to mean something such as "down-river people," was used by Lenape and Jersey Indian descendants to refer to those Lenape who had formerly (before 1740) lived in southeastern Pennsylvania and northern Delaware. *Unalachtigo*, or terms that are variants of that spelling, appears to have referred to the Jersey Indians, based on their entirely independent patterns of activity in the face of colonial activity. Certainly by the 1760s, when Hannah Freeman and other Lenape sought sanctuary in New Jersey, the native population neither rushed to offer aid nor provided any incentives for her to remain (see Becker n.d.c).

Although the cultural patterns, and probably the archaeological record, of the Lenape and their Jersey kin may have been indistinguishable in the main, ties of kinship and other behavioral concerns led these two peoples to trace very different courses through the historical record. The data on social organization and status ranking noted below are believed to be equally applicable to both populations, and one might predict that their material culture as well as belief systems were nearly indistinguishable. Thus, on some level, one could say that these people shared culture. However, how they differed in their implementation of the rules is the concern of this paper.

Social Organization

The evidence for matrilineal descent among the Lenape is clear from data on the earliest land sales, descriptions in the middle of the seventeenth century, and in every relevant document known until long after they left the Delaware Valley, although the basic economic unit appears to have formed the basis for several bands that can be identified. Each band appears to have inhabited a single river valley or stream valley flowing into the Delaware, with the exception of small bands occupying the area of several small streams (e.g., the Okehocking). The nature of bands along the Schuylkill is still unclear, but possibly several small groups inhabited the drainage of this large river feeding into the Delaware. In general, evidence suggests that each of these groups may have been an extended family band, the largest of which may not have included more than fifty men, women, and children.

All of the evidence from the historic literature and the slight archaeological data indicates that the Lenape were an egalitarian people during their entire history along the Lower Delaware River drainage. The classic pattern for foragers appears to apply to these peoples; that is, status is achieved through gender, age, and ephemeral personal characteristics. Thus we find that males consistently appear as grantors of land in land sales, negotiators, and counselors, despite clear use of matrilineal descent. Only on rare occasions are females noted by name. This may reflect the strong patrilineal viewpoint of the Europeans recording the information, but on the whole the fundamental ascription of status by gender appears evident.

Age clearly acts as a major factor in status, with young males always appearing at the ends of lists (such as those lists of grantors found on deeds, or of lists of Lenape attending a treaty) and elder members appearing near the beginnings. The importance of personal characteristics in status ranking arises in this "senior" context since many individuals who appear to be the oldest members of their groups never appear first on these lists. Thus, some demonstration of individual merit, skill, or negotiating ability may give a particular Lenape adult higher status than an older member of the same band.

After the Lenape bands left southeastern Pennsylvania the traditional technique of identifying the extended family coresidential band by a natural feature in their environment began to change. A few of the settlements of Lenape-Delaware after 1750 are identified by the name of an individual whom we may infer was recognized as an informal "leader," such as Custaloga's Town (Kent, Rice, and Ota 1981, 32). This shift toward acknowledgment of a specific leader was extremely gradual and may have occurred less among the Lenape than among their former neighbors who also moved out along the frontier.

Burial data also provide some indications of status systems. During the Late Woodland–Early Contact period, individual burials appear to have been made in shallow holes with bodies in a flexed position. By the period 1720–30 most Lenape individuals were buried in an extended position; however, separate graves, which are regularly spaced, continue to characterize most Lenape burial areas. The mortuary pattern is clearly distinct from those of populations living to the south of the Lenape realm, who appear to have been organized in low-level chiefdoms and to have used secondary (or bundle) burial techniques leading to the formation of large ossuaries. In sum, the ethnohistoric data suggest band-level organizations and flexible social systems. It is interesting to note that the flexibility of Lenape responses may have allowed the survival of these people, with language and customs relatively intact, into the twentieth century.

So far the period under discussion extends from about 1550 to 1750, which is far more than a brief interval within which no culture change could be expected. Almost all the historic data dates from after 1620, when the major disruptions caused by the introduction of steel tools and weapons seem to have had their greatest effect. The seventy years of "initial contact" may have only been the final phase of earlier Lenape movements. How these possible changes in borders influenced Lenape band interaction is not known, but probably there was little effect on the individual Lenape bands and their responses to subsequent colonial expansion. The events described below document two known patterns that relate to colonial interaction with two Lenape bands. The differences discussed are believed to reflect inherent flexibility in band organization, rather than behaviors peculiar to Lenape culture as such. The descriptions provided below are based on two earlier publications (Becker 1976; Becker 1980a).

The Okehocking Band

From about 1670 to 1680 the Okehocking band of the Lenape appear to have resumed traditional use of their ancestral lands. Extended families maintained a summer station on or near the Delaware River during summer and early fall. Family units then extended into the hinterlands to hunt during the winter. Innumerable variations on this pattern have been proposed, but in essence they show the same basic features of foragers adjusted to the local particular environment.

Land sales and colonial trading posts and settlements made little difference to the Lenape lifestyle, except that some intensification of maize agriculture, and possibly settlement aggregation, may have characterized the period from about 1650–70. The traditional extended family units appear to have occupied the various valleys that had been inherited through their respective matrilines. Lenape agricultural surpluses (maize), which had been exchanged for alcohol in the 1660s, seem not to have found a market in the 1670s. Quite possibly colonial farmers were producing crops large enough to insure self-sufficiency by this time.

The political pressures from the Maryland colony around 1670 were turning into military pressures (Jennings 1982) apparently aimed at asserting territorial claims in the southern Lenape area by right of conquest (Becker n.d.c.). This problem was solved when the Crown placed William Penn and his Quakers (Religious Society of Friends) between New York and Maryland. Penn's activities in his colonial empire altered the problem of fixing boundaries, from the Crown's point of view, and gave the Lenape a European ally to replace the Swedes and Dutch. The lesson, which was well known to various New England native groups (that the English came to settle, not only to trade), was not lost on the Lenape. However, the dispersal of the Susquehannock by the Marylanders around 1675 left open a vast territory for hunting and jobs as middlemen in the fur trade. Many Lenape appear to have moved into this niche quite early, using their ability to operate independently to assume this option (Becker n.d.c). The majority of the Lenape, all of whom seem to have sold their lands to Penn, opted to remain on those tracts where they were settled and to which Penn acknowledged their de facto land rights (Becker n.d.f).

This conservative mode of adjustment to European settlement, midway between the complete abandonment of the area and the other extreme of marrying into or becoming peripheral to colonial society, did not prove to be a troublefree course. The considerable immigration of Quakers and other people into the region and their even more remarkable reproductive patterns rapidly filled the river banks with farmers. Within a few years these efficient postmedieval technicians were purchasing land and clearing the forest along every one of the Delaware River tributaries. The Lenape bands uniformly responded by relocating their summer stations further up these feeder streams than any colonial settlement had reached. This process, which seems to have been repeated at fifteen- to twenty-year intervals, correlates to some extent with the periodic relocation of these encampments in the vicinity of the river mouths to provide better access to forest resources and rich soils depleted in the vicinity of their former location. These upstream movements, however, were far from deep and salt water fishing, away from migratory fowl, and often cut off from anadromous fish by the construction of dams by colonists securing water power. Despite these drawbacks, which increased with time and with distance from the Delaware River, the traditionalists among the Lenape could sustain a semblance of their foraging lifestyle from these stressed situations. Certainly as the difficulties increased, more Lenape individuals or families must have departed for the frontier, and more "drop outs" must have affiliated with and lived among the colonials than ever before (see Becker n.d.g).

Around 1700, the Okehocking band, probably then living in the vicinity of present Ridley Creek State Park, petitioned the proprietors for a "secure" tract of land. Although no trespass or other problems appear related to the land on which they were then living, colonial settlement rapidly was encircling them and they certainly felt a need to relocate (Becker 1976). Their petition to the government was granted almost at once. A warrant for a five-hundred-acre grant was written up and a survey conducted soon after. A deed for five-hundred acres such as would be granted any colonial purchaser was drafted. Instead of a stipulated price, the proprietors granted this tract for the exclusive and perpetual use of the Okehocking so long as they saw fit to continue their occupation. Such a grant was in accord with Penn's concern for protecting all Lenape on lands that they had "seated" and had the added advantage of clarifying the boundaries of this particular parcel.

This benefit to the proprietary government was immediately evident because now colonial purchasers knew exactly where the borders of the Okehocking tract lay and could buy land immediately adjacent to it. In the case of other Lenape bands and the Okehocking before 1701, colonials avoided purchasing land near a Lenape settlement or habitation area lest they by accident include some area claimed by the band and therefore unable to be sold. Such a problem would negate the sale and require the colonist to seek out a warrant and wait for a survey of yet another tract, which would result in costly delays.

Since the Brandywine band and others also moved up their respective streams around 1700–1701, one might wonder why the Okehocking were the only ones to petition the government. The valleys occupied by Ridley and Crum creeks are quite small and the watershed area limited. As the Okehocking moved inland away from the Delaware they found that their available territory became smaller since it was encircled by the upper feeders of both the Brandywine and the Schuylkill. In effect, the Okehocking had no place to go. Their options included leaving the area, trying to merge with other bands—which were also having problems, or trying to secure a buffer (boundary line) to avoid the apparent encroachment of the colonists. Perhaps the younger Okehocking fled the area at a rate faster than did the other bands. By 1737 the Okehocking had abandoned this tract forever. By that year, or perhaps two or three later, all of the Lenape bands had left the Delaware Valley, even if only by shifting a few miles to the smaller feeders of the rivers flowing west toward the Susquehanna and toward the sunset.

Given the advantages to the proprietary government of fixing the boundaries of Lenape band, one might wonder why they did not suggest a similar arrangement with other bands. This, however, is not of concern in this paper, which observes the variations in strategies used by different bands sharing the same culture. Another attempt to accommodate to the stresses felt by each of these bands can be described from the Lenape band occupying the stream to the southwest of the Okehocking.

The Brandywine Band

The pattern of withdrawal from the agricultural expansion of colonial settlers described for all Lenape bands applies quite clearly to those who lived along the Brandywine after 1600 (see Becker 1980, 27). This band was probably derived from the group living around Hopokahocking (Fort Christina/Wilmington) when the Swedes arrived in 1638. We do not know whether these "owners" of the land actually resided there during the years of the Susquehannock hostilities, but certainly Swedish purchases of land rights were made from contemporary owners. Although their relocations upstream is not yet known in detail, this band settled at the Big Bend of the Brandywine about 1680 and stayed there until 1701. As the Okehocking were moving to their granted tract of land, the Brandywine band was relocating to an area near present Northbrook, Pennsylvania. They used that area for some twenty years before taking up residence at "The Last Stop," a settlement site (and associated burial area) near Glenmoore. Here they lived until about 1733 when they, like other Lenape bands soon after, relocated to the west, where the streams feed the Susquehanna.

This pattern of relocation-withdrawal appears typical for all Lenape bands, but the Brandywine band did not leave the Northbrook site that quietly. Unlike all of the other Lenape bands, the Brandywine group traveled to Philadelphia to register a claim for the lands along the upper reaches of the Brandywine Creek. They claimed that they had been given a deed that provided them with the title to a strip of land one mile wide on either side of the Brandywine above the large rock of Northbrook. This deed, however, had been destroyed in a fire some years before. This petition led the proprietors to search for anyone who could independently support this petition. Depositions were taken from neighbors and other residents along the Brandywine (Logan Papers: Indian Affairs; Historical Society of Pennsylvania), but nothing solid could be found to confirm this claim.

On the allegation alone, this band, led by the speaker Checochinicon, was given goods with a value sufficient to satisfy their claim. The Brandywine band also is believed to have "sold" their "rights" to lands near Glenmoore to John Henderson, who was a squatter farming a tract on Springton Manor, where this band lived in 1733. In short, these people seem to have found that rewards were to be had for the asking. Although they requested land along the Brandywine by previous deed and the Okehocking previously had been given a "secure tract, no mention of offering a warrant to the Brandywine band is made in any deeds, nor do they seem to have wished one.

Quite possibly these Brandywine people did at one time have a deed referring to a land sale. Such deeds were drawn up as indentures at every land sale, probably back to the 1629 sale. These indentures were the standard contract form, with two copies of the deed (or other document) being written on opposite ends of a parchment. At the signing the pair would be separated by cutting a wavy line at the middle and then appropriate seals could be added. The validity of any claim could be tested by comparing the pieces held by both principals—the matched indentures of the cut proving the validity of the contract. Although the procedure is not often recorded in historic documents, it was so common that everyone knew the formula. The Lenape, although none could read, would have someone read their copy and used their memories to validate the text. The deed recalled by these petitioners could have been a Swedish, Dutch, or early English deed for lands unknown because no one who had seen the alleged deed could recall its contents. (and several colonists who thought that they had seen it were themselves illiterate).

From the vantage point of history and review of all of the known deeds, there does not appear to have been any likelihood that the deed in question, which may have existed, granted strips of land along the Brandywine to these people. More likely, one George Harland, a "friend" to the Brandywine band, may have put them up to making this claim, perhaps alleging that he had known the contents of the deed. Harland had taken advantage of his knowledge of this band's activities when he filed to purchase their recently abandoned land at Big Bend. Quite possibly Harland, known to have been involved in various schemes, thought that he could reserve for the future choice land along the upper Brandywine if his Lenape friends had their claims honored. Should they depart once again, as he most certainly knew they would, he might be able to purchase the land at a favorable rate.

When the Brandywine band pressed this claim they may have believed in all good faith that they had a legitimate right to the land. The response of the government, treating the petition as having the possibility of validity, led to a long and costly search for evidence. Despite the failure of this search, in the best tradition of William Penn, a settlement was made.

Important to the emphasis of this paper is the great difference in the way the Brandywine band went about responding to land stress as compared with that of the Okehocking. Both of these groups, when compared with other Lenape, are unusual in that they attempted to work through the colonial government to secure their ends. In no other cases do we find a similar behavior among the bands in southeastern Pennsylvania. Only the compliants regarding the Walking Purchase Confirmation Treaty (1737), which involves a large amount of land that appears to have been outside the Lenape area, are remotely similar in nature. The principal complainants in 1737 were Jersey Indians attempting to legitimize claims to lands in Pennsylvania that they held only by right of recent residence.

Conclusion

The data from historical documents, supported by limited archaeological evidence, suggest that the peoples of the lower Delaware Valley shared linguistic traits and may be identified as a single culture, that of the Lenape. However, considerable differences existed between the Lenape groups on the east and west sides of the Delaware River. Furthermore, even proximal bands on the Pennsylvania side, occupying nearly adjacent stream valleys feeding into the Delaware River, responded differently and independently to colonial expansion. These differences are by no means a reflection of disorganization and lack of social or political cohesion, as some historians have suggested, but rather as a reflection of expected band independence and flexibility. This flexibility operated on the individual as well as the group level during and before the historic period. Variations in the archaeological record of the terminal Later Woodland period reflected the presence of proto-Lenape people, whose transition into the Contact Period and beyond was successfully met by flexible traditional strategies.

5
Late Woodland Settlement Patterns in the Upper Delaware Valley

HERBERT C. KRAFT

In Late Woodland and early colonial times, the tri-states region above the Delaware Water Gap was the homeland of the Minisink Indians. The earliest known historic reference to this area is by Thomas Budd, who in 1685 wrote, "From the Falls of the Delaware River the Indians go in cannows up the said River to an Indian Town called Minisincks which is accounted from the Falls about eighty miles . . . [and here] there is a great quantity of exceeding rich, open land" (Budd 1685, 30).

The Minisink Indians were part of a much larger but loosely affiliated group that shared a similar culture and spoke the same Algonquian language—a dialect that linguists have termed the Munsee or M-dialect of the Delaware language (Goddard 1974; 1978b, 72–73, 237). As these individual groups came into contact with the Dutch and English settlers, they at first attempted to cooperate, trade with, and otherwise accommodate themselves to European ways; but conflict and hostilities, such as the Dutch massacre of eight peaceful Hackensack Indians at Pavonia (DeVries 1909, 227–29), epidemic diseases, and other conditions forced the Indians gradually to relinquish their traditional lands and withdraw into the interior.

During the 17th and early 18th centuries, the Upper Delaware Valley received many of these remnant peoples. Here they amalgamated with the resident Minisinks and gradually lost their identities as Esopus, Rechgawawanks, Hackensacks, or Raritans. The Europeans simply lumped them all together as Munsees. The term *Munsee* is not an Indian word, although the usage has become so common that decendants of the Munsee-speaking people living in Wisconsin, Ontario, and elsewhere identify themselves as Munsee or Munsee Delawares even today. The name *Munsee* has never been observed in early New Jersey accounts and was first recorded in the Pennsylvania records in 1727. In the early historic documents, *Minsi* is a corruption of Minisink or Minnissinger as the Moravian missionaries recorded it (see Hunter 1974; Hunter 1978; Kraft 1975, 61).

As a consequence of more than a decade of archaeological excavations and research sponsored by the National Park Service, we know more about the Minisink region and the lifeways of these late prehistoric/early historic people than we do about most other Indian people in the central and northern Middle Atlantic region. Not only was funding made available for excavations, but the Upper Delaware Valley itself was less disturbed and less built upon than other areas. There was, therefore, a better chance of finding house patterns, burials, refuse and storage pits, and associated artifacts than would have been the case in the intensively urbanized and industrialized cities of, for example, Hudson and Bergen counties in northern New Jersey or in New York City, Camden, and Trenton. In this paper the term *Upper Delaware Valley* refers to both the New Jersey and Pennsylvania sides of the Delaware River above the Delaware Water Gap and extending north beyond Port Jervis, New York. The Late Woodland or Horticultural period in the Upper Delaware Valley, which is the subject of this paper, began about A.D. 1000 and continued into the historic period when the Euro-American settlers finally evicted the native peoples from their traditional homelands.

Archaeological and Cultural Terminology

In 1948, Dr. William A. Ritchie excavated on the Bell Philhower site and thereupon introduced New York State archaeological terms into the Upper Delaware Valley, such as Owasco, Castle Creek focus, Chance phase, and the like (Ritchie 1949). Following several seasons of my own extensive excavations in the late 1960s and 1970s, I found it necessary to reconsider this terminology with respect to its applicability and appropriateness to the Minisink area. By 1975 I had become convinced that the historic Minisink Indians had lived in this same valley for many centuries and, from all indications, had emerged from people who had lived here at a still earlier time. In other words, there appeared to be an *in situ* development of the Minisink Indians. On the other hand, the name *Owasco,* as used by Ritchie, really identifies a pre-Iroquoian people, whose heartland was the Finger Lakes and Mohawk River–St. Lawrence River drainage (Ritchie 1969, 272–302). *Owasco* identifies archaeological assemblages associated with people who, from all indications, spoke an Iroquoian language and lived in a relatively complex sociopolitical setting. In contrast, the Minisink people spoke an Algonquian language and lived in small, sometimes individual groups or bands, and not in villages with or without fortifications, as did the Owasco groups.

Nonetheless, there are some striking similarities between the

Owasco-Iroquois culture and the Upper Delaware Valley peoples. The ceramic pots are very much alike and so too are certain of the stone artifacts. Similarities in pottery, however, do not necessarily presuppose similarities in people, a similar language, or similar culture (cf. Griffin 1978, 272). Moreover, the radiocarbon dates on pottery-bearing sites in the Upper Delaware Valley are coeval with those in upper New York State. We cannot, therefore, be sure which way the influence went. It is possible that culture borrowing could as easily have gone from the Upper Delaware Valley into New York State as vice versa, the way Ritchie has suggested. I prefer to think that there was a somewhat friendly interchange of cultures between these two areas before the fur trade and European-Indian alliances strained native American intergroup relationships. I presented in 1975 further arguments against this use of northern New York State terms to identify the patently Algonquian people of the tri-states area and the northern New Jersey and Metropolitan New York area (1975a; 59–61).

To avoid confusion in archaeological matters pertaining to the upper Delaware Valley and contiguous areas, the term *Pahaquarra phase* here replaces *Owasco* when I refer to the people who lived in the Upper Delaware Valley during the early part of the Late Woodland period. The culture of the hunting, fishing, gathering, and gardening people of the Pahaquarra phase is most easily distinguished by the collarless, cord-decorated, Owasco-style pottery (Kraft 1975b, 119–32; 1975a, 101–20). The name *Pahaquarra* is derived from the township that extends from the Delaware Water Gap to the Wallpack Bend along the New Jersey side of the Delaware River in Warren County. Specifically, the Pahaquarra phase is an archaeological designation that refers to the ancestors of the Minisink Indians in the time period of about A.D. 1000 to 1350. The Minisink phase (A.D. 1300–1700, which succeeds the Pahaquarra phase in the Upper Delaware Valley, recognizes the prehistoric tradition that continued into the historic times marked by Indian/European contact. The way of life did not change significantly. Minisink, or Minnissinke and other variations (see Goddard 1974, 237), and the archaeological Minisink phase identify the same people who lived in the Upper Delaware Valley before Euro-American colonization. This Late Woodland and protohistoric culture is distinguished by certain incised, medium- to high-collar ceramic vessels (Kraft 1975a, 134–49; 1975b, 120–36), by the use of round-ended, dome-shaped, bark-covered longhouses of small to medium size (about 18 to 60 feet long), and by a distinctive array of domestic and procurement implements including the triangular arrowhead. *Munsee* is the historic Euro-American term for the Minisink Indians and those remnant bands of Munsee-speaking Indians who joined the Minisinks in their westward and northward migrations out of New Jersey in the eighteenth century.

It has been noted (Goddard 1978b) that the Minisink Indians and certain other historic bands such as the Canarsee, Esopus, Hackensack, Haverstraw, Kichtawank, Matinecock, Massapequa, Navasink, Nochpeem, Raritan, Rechgawawank, Rockaway, Sinsink, Tappan, Wappinger, Warranawankong, Wiechquaeeskeck, as well as the Minisink Indians and other unnamed peole in northwestern New Jersey, northeastern Pennsylvania, and southeastern New York State, shared a common Munsee dialect of the Algonquian Delaware language. In this language and in certain material and spiritual/cultural attributes they differed from the Unami-speaking people who lived in central and southern New Jersey, eastern Pennsylvania, and Delaware. Among the Unami or U-dialect speakers were the Armewamex, Atsayonck, Big and Little Siconese, Mantaes, Naraticonck, Okehocking, Remkokes, Sanhikan, Schuylkill, and Sewapois, for example (Goddard 1978b, 238; Weslager 1954). Although these numerous bands and/or geographical assemblages were independent and frequently fluid in their constituency, they nonetheless shared certain cultural characteristics and certain vaguely defined allegiances. In the archaeological record these peoples are identified by prevailingly collarless pottery forms and designs, in different burial orientations, and in slightly different tool kits from those of the Munsee. The linguist perceives dialectical differences that even today distinguish the Unami-speaking Delawares from the Munsee-speakers. In order to deal collectively with these two cultural/linguistic groups, and for the sake of convenience, it has become necessary to establish and define several existing and contrived terms such as Munsee and proto-Munsee, Unami and proto-Unami, Lenape and Delaware.

Munsee, as already noted, here refers only to those northern Munsee-speaking people who amalgamated with the Minisink Indians and ultimately emigrated to New York State, Wisconsin, and Ontario, Canada, where most of their descendants are living today. *Proto-Munsee* refers to this collective group as we recognized them in late prehistoric and early historic times, while they were still living in their respective home territories. At the Treaty of Easton in 1758, this collective Munsee-speaking group was recognized as living north of an east-west line drawn from

> . . . the mouth of the Rariton [sic] up to Alametung Falls in the north branch of the Rariton river, thence on a streight line to Paoqualin

Mountain [the Delaware Water Gap] where it joins on Delaware River (S. Smith 1890, 493).

Unami, another historic term, first employed in the official Pennsylvania records in 1757 (Hunter 1974; 1978), is restricted to those Unami-speaking people who emigrated from their traditional homeland and whose descendants are now living, for the most part, in Oklahoma. The term *proto-Unami* identifies these collective historic Unami-speaking bands while they were still living in their traditional homeland below the line established at the Treaty of Easton.

The name *Delaware,* which now commonly describes both of these groups, is in fact an English name for Sir Thomas West, Lord De La Warre, first governor of Virginia. *Lenape* is a Unami self-designation meaning, among other things, "a male of our kind" or "our men" (Brinton 1885, 35); "ordinary" or "real" or "original" people (Goddard 1978a, 235); or "common people" (Nora T. Dean, personal communication). This term has been used for more than a century to refer to all of the people who spoke the Algonquian Delaware tongue, the Unami- as well as the Munsee-speakers, who lived in the geographical region that now comprises the entire state of New Jersey, southern New York State north to Kingston, and western Long Island, eastern Pennsylvania, and the northern sector of the State of Delaware. There being no other suitable name, *Lenape* will designate all of the native people who in protohistoric times occupied this geographical region.

Environment and Subsistence

The Delaware River and its tributary streams, the lakes and ponds, the fertile alluvial floodplains, and the wooded Kittatinny Mountains and Pocono foothills provided a rich environmental setting. The Delaware River permitted easy canoe transportation, but perhaps more important, it was a rich ecological resource. In the spring, when the water temperature reached 40 degrees Fahrenheit, the shad began to ascend this river to spawn as far north as present-day Hancock and Deposit, New York, a distance of more than 350 miles from the ocean (Bishop 1935). Early records maintained by the U.S. Fish and Wildlife Service indicate that in the 1880s, and before pollution, some 14 to 20 million pounds of shad were caught annually in the Delaware River (Sykes and Lehman 1957; Mansueti and Koeb 1953; Walberg and Nichols 1967; Chittenden 1972; Miller, Friedusdorff, and Mears 1974, 105–6). Other anadromous fish included sturgeon that attained lengths of over six feet and weighed up to 150 pounds (NJMR 1905, 84–87). Perch, pickerel, muskellung, sunfish, suckers, catfish, and others could be caught all year, thus providing a dependable food resource.

The importance of fishing in the lives of these peoples is attested to by the thousands of notched flat pebble and trimmed rectangular netsinkers that have been found scattered along the shores, and that occur in almost every archaeological excavation in the valley. Of particular interest are the caches of netsinkers, consisting of thirty or more similarly sized and shaped weights, which appear to be all that remains of gathered or folded nets that may have been lost, discarded, or abandoned in times of flooding. The collections of small, medium, or large netsinkers that characterize such caches suggest different applications, as for example, casting nets, seines, or gill nets (Kraft 1975a, 111–18). The lengths of such nets is now uncertain because we do not know how closely the sinkers were spaced along the bottoms of the nets. De Vries (1909, 222) notes that the Indians sometimes used "seines from seventy to eighty fathoms in length, which they braid themselves, and on which, in place of lead they hang stones, and instead of the corks . . . to float them they fasten small sticks of an ell in length." Fish were also caught by means of stone weirs. Few of these weirs remain, but one of stone still located above Dingman's Bridge consists of a wall of large cobbles and boulders laid diagonally from the Pennsylvania side nearly to the New Jersey side, where presumably the Indians had implanted wooden stakes close together to form an enclosure into which the fish could be gathered and then removed by hand or spearing.

Additional archaeological evidence for fishing comes from the extensive deposits of fish scales, opercula, and fish bones found in refuse pits. Although bone harpoons have been found on Owasco and Iroquois sites in New York State and also at the Abbott Farm site near Trenton (Cross 1956, plate 29), no such implement has yet been excavated in the Upper Delaware Valley; nor for that matter, have we yet found fishhooks or fish gorges, although we suspect that they were used.

The Delaware River and its feeder streams also provided the Indians with an abundance of the freshwater mussel *Elliptio complanatus*. Refuse pits containing thousands of such shells are common (Kraft 1975a, 71–73, 156–57). Because of the rich resources of the Delaware River shore, the terraces above the Delaware River were the favored places of Indian habitation, despite the occasional overbank floodings. It was there that they erected oval or round-ended, bark-covered longhouses and planted nearby gardens.

A short distance beyond the Delaware River and the terraced floodplains are the wooded Kittatinny Mountains on the New Jersey side, and the Pocono Mountain foothills on the Pennsylvania side. These areas provided a third environment and a very important ecological resource. It was in these forested slopes that the Indians obtained the saplings, bark, and bast required for house construction, as well as wood for dugout canoes, bowls, bows, and for other manufacturing needs, as well as firewood. These forests sheltered numerous game animals, in addition to providing edible nuts, roots, berries, and other foods for the Minisink Indians. Chestnuts, which today are quite scarce due to the Asiatic blight, must have been very common in prehistoric and early colonial times (De Vries 1909, 219; Zeisberger 1910, 47) and were no doubt gathered in great quantities for winter storage. Other forest foods that are mentioned in the early accounts include wild plums, wild cherries, persimmons, hazelnuts, butternuts, wild grapes, blackberries, and strawberries.

Both deer and elk were also abundant, and their remains are frequently encountered in archaeological refuse pits. Bear too were killed, especially in the winter. The remains of black bear constituted 2.3 percent of the minimum number of individual animals recovered from refuse pits on the Minisink site (Kraft 1978, 29). Turkeys and other forest creatures were snared, trapped, and hunted with bow and arrow. The turkey, in addition to providing meat, was hunted for its feathers, which were fashioned into warm, decorative mantles and quilts (Lindestrom 1925, 221–22). The most important resource of the forest, however, may have been firewood both for cooking and for warmth in the wintertime.

Settlement and Community Patterning

Fourteen house patterns have now been observed on four riverine sites: Miller Field (three houses), Harry's Farm (five houses), Pahaquarra (four houses), Minisink (one house) (Kraft 1970a, 4, 9; Kraft 1970b; Kraft 1975a, 75–86; Kraft 1976a, 64–65, 84–86; Kraft 1978, 22), and on one inland lake location, the Swartswood Lake site, which yielded two house patterns. These Minisink house patterns vary in size from small, oval structures 15 feet in length by 12 feet at the greatest width, to one round-ended longhouse that measured 60 feet long by 20 feet wide.

Most of these houses appear to have been well constructed. The postmolds indicate that pointed saplings measuring about $2\frac{1}{2}$–$3\frac{1}{2}$ inches in diameter were drilled into the earth to a depth of 12 to 16 inches and spaced from 12 to 14 inches apart. A round-ended, dome-shaped inner frame was probably constructed on these saplings with a door frame on one of the long sides. Chestnut, elm, linden, or other types of bark shingles were then attached to this support structure, presumably with bast fiber ties. When the overlapping bark sheets were in place, a second set of posts was sometimes driven into the earth approximately parallel to those already constituting the inner frame, and close against the outer bark covering. These external posts and the attached crossmembers not only reinforced the structure, but they presumably served to clamp and tie the bark sheets, and so hold them in place even in blustery weather.

Minisink Indian longhouses differ from those of the Iroquois in several ways. Minisink houses are, without exception, round-ended longhouses with a single doorway on one of the long sides. The partitions in the larger bark lodges extend from one wall to within 3 or 4 feet of the opposite side, thereby providing a passageway along the entrance side of the structure. Benches or sleeping platforms were placed along one wall between the partitions, which presumably marked off individual family quarters. Storage pits or cellars were dug into the earth at one or both ends of the house (see Kraft 1970b). These pits may have contained dried foods such as nuts, berries, maize, beans, dried meat, or fish for easy access during inclement weather, or in the winter when snow covered the outside storage pits. At least one house had a small shedlike attachment, suggesting a granary or possible woodshed (Kraft 1976a, 85).

In 1680, the Labadist ministers Jasper Dankers and Peter Sluyter described a Nyack Indian house that is probably similar to the Upper Delaware archaeological features. The great number of people who occupied such an abode is evident:

> .... We went... to her habitation, where we found the whole troop together, consisting of seven or eight families, and twenty or twenty-two persons ... [in a low house] about sixty feet long and fourteen or fifteen feet wide. The bottom was earth, the sides and roof were made of reeds and the bark of the chestnut trees (Danker and Sluyer 1867, 124).

Johan de Laet also describes a house, possibly in the Mohican area, as "... well constructed of oak bark, and circular in shape, with the appearance of having a vaulted ceiling." He further states that some of these dwellings housed forty men and seventeen women (Laet 1909, 49).

Some later writers have been deceived by map illustrations allegedly showing Minisink houses and villages (Philhower 1953, 1).

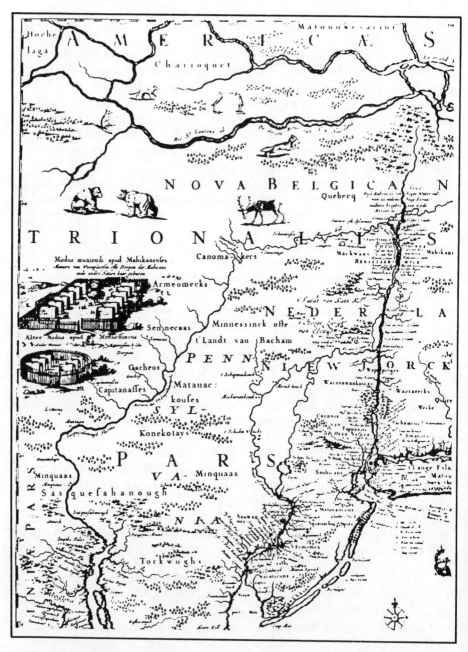

Fig. 12. Detail of Hugo Allard's map of 1673. This copy of the Nicholas Visscher map of 1656 contains the statement "Alter Modus apud Minnesincos. Ander Manier der Minnssincksche Dorpen" between the two stockaded towns.

They cite as evidence John Allard's 1673 reproduction of the earlier Nicholas Visscher (1656) map, which contains the inscription in Latin and Dutch, *"Alter modus apud Minnesincos. Ander Manier der Minnessincksche Dorpen"* (another manner of Minisink villages). This statement is penned between two stockaded Indian villages—one a square enclosure with twelve square-ended, dome-shaped houses of equal size, the other a circular stockade enclosing six similar houses (fig. 12). These cartographers' devices were apparently modeled after the 1585–87 John White drawing of Indian villages in North Carolina and Virginia, which were later reproduced by Theodore De Bry (Lorant 1946, 190–91). Such fillers, or ornaments, were often used by early mapmakers, who wished to cover up areas where little or nothing was known, and for such purposes the illustrations of deer, bear, Indians, and Indian villages were both interesting and attractive. However, such engravings cannot be considered as eye-witness documentation of what Indian villages really looked like. Kraft (1977, 6–19) provides a more detailed discussion of this problem with respect to the Minisinks.

Not one archaeological site in the Upper Delaware Valley has ever produced evidence of a stockaded or fortified Indian village, and only rarely are more than two such houses in close proximity. But even then we cannot be absolutely sure that such houses were standing together at the same time. On the contrary, all evidence recovered thus far suggests that these people lived in small family units or farmsteads probably consisting of extended related families. There is no evidence that these Minisinks had any fear of attack; hence, there was no need for fortification.

Another interesting aspect of settlement and community patterning is the large complex of storage features associated with house patterns at most sites. The refuse pits or filled-up storage pits from archaeological sites in the Minisink area vary both in shape and size. Soil conditions certainly determined how large or deep such pits might be dug. The riverine slit on most sites is very stable and structurally supportive. On the Miller Field, Harry's Farm, and Pahaquarra sites, for example, the pits commonly measured 3 feet in diameter and 2 or 3 feet deep. Occasional deep silo-shaped storage pits measured up to 5 feet in diameter and up to 93 inches below the interface with the plow zone. The faint zigzag outlines of the pit walls suggest that they were dug with wooden dibbles. The soil was probably removed in baskets and the walls of the pits were then troweled smooth with the use of teshoas (crude chipped stone flake tools), many of which were found in or proximate to such refuse pits. In order to make such storage pits more impervious to ground water seepage, the walls were sometimes puddled with a thin layer of clay. Postmolds in the bottoms of these large storage cellars suggest that ladders, or more probably racks or roof supports, were implanted in the pit floors (Kraft 1975a, 67).

Deep storage pits doubtless served for a number of years. Before each successive reuse, however, the Indians appear to have burned dried corncobs, bark, or wood inside the larger pits with the apparent intention of killing off the insects and vermin that had fallen into the pit, and also to bake and reharden the walls. Such charred corncobs have been preserved in a number of silo-shaped pits, and together with spilled beans, squash seeds, nuts, and other floral remains, they provide verifiable information concerning the foods these peoples raised, collected, and stored.

The gardens that supplied the cultigens were apparently located close to the houses. Evidence for horticulture comes not only from the charred remains of maize, beans, and cucurbits, but also from the stone hoes found in limited numbers on Late Woodland sites. If the scapulas of deer or elk were ever used as hoes in the Minisink area, the evidence has now decayed without trace. Other domestic implements include roller pestles for use in tree trunk mortars and mullers and millingstones.

The teshoa, already mentioned as a pit wall-smoother, was a ubiquitous Late Woodland tool in the Minisink area. Consisting merely of a larger spall struck from a river cobble, it could be used as a knife, scrapper, chopper, and even whetstone depending upon application (Kraft 1966; Kraft 1975a, 102–6). Another kind of throw-away knife, called a ground slab knife (Kraft 1975a, 106), consists of a natural slab of traprock, slate, or similar stone whose edge was ground or honed until both sides converged, celtlike, to a cutting edge.

An interesting tool phenomenon, observed on several sites, was the use of a triangular point, apparently placed tip into handle so that the straight base could be used as a makeshift scraper or strike-a-light (Kraft 1975a, fig. 65h, i, 92). Such points were occasionally also pressed into service as drills, reamers, and knives.

The sturdily constructed house sites of the Minisink Indians and their associated storage features were fairly permanent year-round settlement sites. This is not to say that members of the family would not leave for extended periods in the fall or winter to hunt and gather in the interior regions. At such times some members of the family— the very old or women with children—may have remained. The small individual or extended family farmsteads of the Minisink Indians were not clustered into villages; instead, they were strung out like beads on a rosary along the fertile alluvial terraces that run like a ribbon between the water and the mountain slopes on each side of the

Delaware River. In less confined areas, however, the settlement may have been arranged in a different manner.

Burial Data and Ideology

Indians of the Late Woodland times used several methods of burial. Cremation, a common practice in Early Woodland times on Adena-related sites such as the Rosenkrans site (Kraft 1976b), was no longer practiced. Inhumation in a flexed position with arms crossed over the chest or hands near the head, and knees more or less tightly bent as in a sleeping posture, referred to as the "sitting position" in some early accounts (Goddard 1978a, 219), is the usual mode of burial. By early Historic times, around circa 1700–50, burial in an extended posture, possibly in imitation of European interments, was also being practiced among some of the Minisink Indians (Kraft 1976a, 55–57; 1978a, 51–52).

Bundle burials, or reburials, are particularly interesting. Several examples have been excavated in which the individual had presumably died at some distance from the family settlement, thus necessitating temporary burial or exposure in a charnel house. Later, the skeletal remains were separated into head, sometimes articulated rib cage, long bones and other small bones, which were then gathered into a bundle for convenient transport to the permanent family campsite or to a new settlement, where the individual was reburied (Kraft 1978a, 51, 53). One especially informative reburial consisting of the skull of a six-to-seven-year-old child has been interpreted as a reburial with evidence of a "feast of the dead" (Kraft 1974b, 29; 1976a, 50–53). The skull of this child, but with no other post cranial bones, was deposited in a pit laden with the remains of numerous animals, birds, fish, and shellfish, together with a decorated turtle shell carapace dish and six bone skewer eating utensils. Other objects from this pit consisted of cut and trimmed turkey bone beads and a fragment of a bone flute. There is absolutely no evidence of cannibalism; rather, the coarse sand and a small left-handed whelk found within the cranium of this child suggest an original burial in a coastal area.

Few Late Woodland burials in the Upper Delaware Valley have associated grave goods. Two burials, both from the Pahaquarra site, were accompanied by entire pottery vessels, which presumably once contained food, and a bone skewer. Tobacco pipes are occasionally encountered with burials of both sexes, and in one instance, on the Miller Field site, a male was buried with a tobbaco pipe and a celt bearing an effigy face (Kraft 1980). The unaccompanied burials may have had perishable offerings placed into their graves such as meat, plant foods wrapped in bark, wooden bowls or baskets, or objects made of skin or feathers, all of which have disintegrated without a trace.

In Historic or Contact period times trade materials of European manufacture began to accompany some of the burials (Heye and Pepper 1915; Cross 1941, 113). The most spectacular Contact period burials derived from the Pahaquarra site were an adult male and female and a child interred in separate graves, accompanied by mortuary gifts. The man and woman were wrapped in bark, the child was placed into a nailed pentagonal pine box coffin. The man cradled the remains of a nonfunctioning musket in his right arm. This gun, lacking butt and trigger guard and with a broken lockspring, has now been identified (by Dr. Donald Baird of the Princeton University Museum) as made by Wilson of London, who began gunsmithing in 1730. A King George I medallion, found among the several necklaces worn by the woman, provided a corroborative date of circa 1726–29. The considerable wear on the medallion and the abuse sustained by the gun would indicate prolonged use; hence a date of about 1750 is postulated for these burials (Kraft 1974a, 45; 1976a, 56–63).

Historic trade items, however, are comparatively rare, not only in the Upper Delaware Valley but throughout New Jersey. In fact, there were substantially more trade items found in the Minisink area than have been reported from any other region of New Jersey. By comparison with the Seneca area of New York State or the Susquehanna Valley of Pennsylvania, however, this sample is meager indeed. The gun discussed above is the only gun ever found in a New Jersey Indian grave, and there are only two known iron axes, both from the Minisink site, two or possibly three brass kettles, perhaps five hoes, three or four seal-end spoons, remains of two glass bottles, and a quantity of glass beads and assorted small metal objects (Heye and Pepper 1915; Kraft 1974a, 47–48; 1975a, 152–55; 1976a, 56–67).

Preferred burial orientations separate the Minisink and presumably other proto-Munsee burials from the proto-Unami burials of central and southern New Jersey, southeastern Pennsylvania, and northern Delaware. The Minisink burials show a decided preference for orienting the head of the deceased in a westerly or southwesterly direction (Kraft 1975a, 86–91; 1976a, 47–63; 1978, 47–55), whereas the proto-Unami burials are usually given eastern or northeastern orientation (Cross 1941, 110–11, passim; 1956, 60; Becker n.d.h).

One aspect of Minisink Indian life that deserves special mention is the evidence of magico-religious rites and beliefs. Nowhere in the entire Lenape territory have there been found so many effigy faces on pottery vessels, on cobblestones, on pendants, and on tobacco pipes. The countenance is simple: two eyes and a mouth. As such it probably represents the *mesingw* or Living Solid Face, the manito who cares for animals of the forest and simultaneously ensures that humankind has enough to eat (Goddard 1978a). The stark effigy faces are very similar to the faces of the Mesinow carved into the posts that formed the historic Big House Church (Weslager 1972, 13). Although traditional Delaware or Munsee Indians cannot confidently connect the archaeological effigy faces with a Big House that had its florescence and early-twentieth-century demise in Oklahoma, the likelihood of the latter having had its origins in the East is certainly strengthened by these discoveries. One pecked cobblestone effigy face from the Minisink site, found as a deposit within a Kelso corded-rim shard in a charcoal bed, had a radiocarbon dating of A.D. 1380 ± 55 years (Kraft 1976a, 80).

If suspended and hung upside-down, the pendant effigy faces suggest a very private communion between deity and wearer. When one looks down at such a pendant, the face looks back at the beholder. The tobacco pipes with an effigy face at the back of the bowl are another case in point (Kraft 1975c). One "bushy head" pendant found on the Miller Field site strongly suggests an historic cornhusk mask (Kinsey 1972, 53; Kraft 1972). Made of fine-grained sandstone, this upside-down pendant has what appears to be braiding around the mouth, face, and forehead, with fringes radiating around the perimeter. Other magico-religious objects of Minisink times include a petroglyph knife (Kraft 1974c) and a large petroglyph of uncertain age (Kraft 1965, 1969).

In conclusion, analysis of archaeological and ethnohistoric data allows us to understand much more about the prehistoric lifeways of the Upper Delaware Valley during late prehistoric times than was possible heretofore.

6
Late Woodland Cultures of the Lower and Middle Susquehanna Valley

JAY F. CUSTER

This paper provides a review of the Late Woodland cultures of the Middle and Lower Susquehanna Valley. I will discuss the Shenks Ferry and Susquehannock archaeological complexes and describe their settlement/subsistence systems, community organizations, and relative socio-cultural complexity. I have studied both published and unpublished sources and have tried to emphasize unpublished papers, reports, and data that have not been considered in earlier studies of the region. In the Susquehannock case, I use only subsistence and community patterning data apparent from the archaeological record in deference to Barry Kent's more complete study (1984) of the historical and archaeological data on Susquehannock culture.

While reviewing the archaeological data, I realized that in many—if not most—cases, "Late Woodland cultures" have been identified primarily on the basis of similar ceramics and little else. Indeed, in many cases changes in ceramic distributions are equated with the expansion and contraction of specific cultural groups and are used to infer levels of sociocultural complexity (e.g., Clark 1977, 1980). Potter (1980, 1982), Puniello (1980), and others (Griffith 1977; Griffith and Custer n.d.; Hatch 1980, 301–5) have specifically addressed the problems with equating ceramic types with social groups or "archaeological cultures," but a different approach will be used here. In this paper, similarities in Late Woodland ceramics, especially in terms of design motifs and design composition (Shepard 1954), are considered indica-

---

I thank Fred Kinsey for allowing me to study collections and use research files from the North Museum, Franklin and Marshall College. I also thank Charlie Holzinger for the use of his unpublished notes on the Nace site and for sharing his thoughts about the site. Barry Kent has also provided important insights to the Late Woodland period in the Susquehanna Valley and permitted my use of the collections at the William Penn Memorial Museum, Harrisburg, Pennsylvania, for which I am grateful. Finally, I thank Arthur Futer for encouraging my studies of Lancaster County archaeology and for providing unlimited access to his collections. Nonetheless, I am solely responsible for any errors of fact or interpretation in this paper.

tions of more common patterns of social interaction, following the work of Plog (1980). In other words, the spatial distribution of a given series of similar ceramics can be equated with a series of social units that interact more commonly with one another than with other social groups who do not share similar ceramic types. These distributions may or may not match distributions of languages, common adaptations, and social organizations, or other components of "archaeological cultures" (G. R. Willey and Phillips 1958, 47–48). Consequently, common features of community organization, subsistence, and relative social complexity will define varied cultures in the Susquehanna Valley during Late Woodland times.

The traditional view of late Middle Woodland and early Late Woodland archaeology suggests that the poorly defined Clemson Island complex is the temporal precursor to the Shenks Ferry complex (Kent, Smith, and McCann 1971, 331–32; Kinsey and Graybill 1971, 43). However, as noted by Turnbaugh (1977, 209), the term *Clemson Island* has lost much of its validity as a ceramic category and an archaeological complex because the Clemson Island ceramic series has come to include all ceramics that are not Vinette I, Shenks Ferry, or Susquehannock. Nevertheless, recent studies by Hatch (1980, 301–5) have begun to clarify the relationships in the West Branch Valley area and document a series of intermediate decorative motifs between Clemson Island/Owasco and Shenks Ferry ceramics. Hatch's study is especially important because it demonstrates that there is no one-for-one correspondence between ceramic types and social groups or "cultures." Working from this position makes it somewhat easier to consider the temporal antecedents of the Shenks Ferry complex.

For the Middle Susquehanna Valley, there is a basis for recognizing a late Middle Woodland–early Late Woodland Clemson Island complex. Excavations at the Clemson and Book mounds (Jones 1931) revealed a series of disarticulated bundle burials with only a few associated artifacts including "classic" Clemson Island ceramics. The Wells site, a Clemson Island site in Bredford County, has produced radiocarbon dates of circa A.D. 950–A.D. 1100 (Kent, Smith, and McCann 1971, 332). In general, the Clemson Island ceramics show definite similarities to Ritchie's (1965, 273–74; Ritchie and Funk 1973, 187) Hunters Home complex, which is transitional between the Middle Woodland Point Peninsula complex and Late Woodland Owasco complex (Schmitt 1952, 61). In a similar vein, McCann's 1971 analysis of Clemson Island ceramics notes marked similarities in ceramic design motifs between Clemson Island and Owasco ceramics. These similarities are underscored by the cooccurrence of these two ceramic types in the stratified midden at the Fisher Farm site (Hatch 1980,

301–5), as well as at Brock Village (Turnbaugh 1977, 227), Bull Run (Bressler 1980), and the Greys Run Rockshelter sites (Bressler 1975, 227) in the West Branch Valley and other similarly dated sites in New York State (Ritchie 1965, 189). In addition to ceramics, the Clemson Island mortuary ceremonialism at Clemson and Book mounds (Jones 1931) in the Middle Susquehanna Valley and the Brock mound in the West Branch Valley (Turnbaugh 1977, 217–22; Carpenter 1949) show similarities to early Late Woodland complexes of New York. Turnbaugh (1977, 218) notes similarities to Kipp Island No. 4 (Ritchie 1965, 261–65); moreover, there are certain similarities to the late Middle Woodland mortuary complexes of the central Delmarva region (Thomas and Warren 1970a).

Thus, the Clemson Island complex represents an initial Late Woodland occupation of the Middle and Upper Susquehanna Valley that has its greatest degree of interaction with social groups to the west and north. Very little can be said about the community patterning or social organization of these groups except to note that similarly dated sites in New York (Ritchie 1965), the Brock site in the West Branch Valley (Turnbaugh 1977,22–24), and the Fisher Farm site (Hatch and Stevenson 1980) minimally seem to represent hamlets that are defined archaeologically as a cluster of domestic debris that cannot be subdivided into redundant functional units (Fuller 1976). These artifact patterns would correspond to habitation sites composed of limited sets of nuclear families. There are some indications of cultigens from New York sites (Ritchie 1965, 189) and at the Sheep Rock Shelter (Cutler and Blake 1967; Steffy 1968) including maize and squash, and extensive seed plant utilization is also suggested by the Fisher Farm data (L. M. Willey 1980; Baker 1980). The presence of distinct cemeteries with super structural mounds, secondary burial treatments, and limited mortuary ceremonialism also indicates the presence of some multicommunity organizations to support mortuary rituals and mound construction. Similar organizations are hypothesized for Middle-Late Woodland mound complexes in the Ridge and Valley area of western Maryland (R. M. Stewart 1981a).

Although Kinsey and Graybill (1971, 43) note that the Clemson Island "culture" is the most likely precursor to the Shenks Ferry complex in the Lower Susquehanna Valley, Kent, Smith, and McCann (1971, 331) and I. F. Smith (1978, 7) note that these ceramics are not generally found south of Harrisburg in the Susquehanna Valley. At the Nace site (Holzinger 1970), which is located in Washington Boro and which is probably a portion of the early Shenks Ferry Blue Rock site (Heisey and Witmer 1964), a variety of Owasco cordedneck, Bainbridge incised, Levana cord-on-cord, and Kelso corded ceramics are present, in addition to a major Blue Rock Phase Shenks Ferry ceramic assemblage. Only a few shards of Clemson Island ceramics have been identified and although the Owasco and other northern types are present at other sites, along with Clemson Island ceramics, the Nace site provides quite a different association. Therefore, it is necessary to consider other possible temporal antecedents of Shenks Ferry complexes in the Lower Susquehanna Valley.

Middle Woodland ceramics of the Lower Susquehanna Valley include two basic varieties: the Mockley series and a variety of grit- and schist-tempered wares. Mockley ceramics have been found at a series of river island sites in Lancaster and Dauphin counties (I. F. Smith 1978, 30–33), as well as at the Erb Rockshelter (Kent and Packard 1969, 35–36) and at the Conowingo site (McNamara 1982a; 17), and are found primarily within the main river valley. The variety of grit and mica schist-tempered ceramics termed Woodland Cordmarked, advanced mica schist-tempered—exterior cordmarked, and advanced interior—exterior cordmarked by I. F. Smith (1978, 33–36, 38) are all very similar to late Middle Woodland Hell Island ceramics of the Coastal Plain and Piedmont Upland areas (Griffith 1982). The shell-tempered Mockley ceramics are not at all similar to Shenks Ferry ceramics and most likely indicate interaction and exchange networks that extend southward along the main branch of the Susquehanna into the Upper Chesapeake Bay (Custer, McNamara, and Ward n.d.). On the other hand, the Hell Island wares and similar types are quite the same in paste, temper, and construction to Clemson Island, Owascoid, and Shenks Ferry ceramics (Heisey 1971, 46–47), and have a distribution that includes both the main river valley and the interior areas of the Piedmont Uplands (Custer, McNamara, and Ward n.d.). Therefore, the basic ceramic technologies of the local Middle Woodland societies of the Lower Susquehanna Valley provide a basis for later ceramic technological developments and show group interaction patterns that extend both to the south and east and to the north. We know very little about these transitional Middle/Late Woodland societies except that within the Piedmont Uplands and High Coastal Plain Middle Woodland societies and the local Late Woodland Minguannan complex, groups lack sedentary communities of any kind and show nonhorticultural subsistence systems very similar to those of late Archaic and Early Woodland times (Custer and Wallace 1982).

In figure 13 are the distributions of late Middle Woodland and early Late Woodland ceramics in the study area and the trends in regional intergroup interaction networks, as indicated by similarities in ceramic technologies and styles. Groups in the Middle Susquehanna

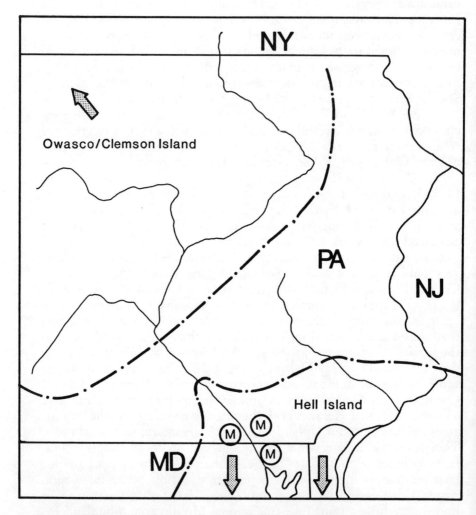

Fig. 13. Middle–Late Woodland ceramic distributions and interaction networks. M-Mockley ceramics.

Valley seem to interact more commonly with groups associated with Owasco complexes to the north and west, while Lower Susquehanna Valley groups interact more often with groups to the south and east. Nevertheless, the basic similarities in ceramic technologies between the Middle and Lower Susquehanna Valley Middle Woodland non-shell-tempered ceramics suggest that the ultimate origins of grit- and schist-tempered ceramics are in the northern areas of the Susquehanna Valley. Similarly, the Middle Susquehanna Valley groups show more complex social organizations than those of the Lower Valley and also at least partially rely on cultivated plant foods. In spite of these differences, by A.D. 1200 these diverse societies became linked within a single interaction network that has been labeled the Shenks Ferry complex.

Shenks Ferry Complex

The type site for the Shenks Ferry complex, excavated in the Lower Susquehanna Valley by Donald Cadzow in 1930, was initially recognized as a late prehistoric "Algonkian culture" site (Cadzow 1936, 61); however, the regional significance of the Shenks Ferry complex was not recognized until Witthoft noted similarities among ceramics from Cadzow's excavations and ceramics from the Summy and Miller sites in Lebanon County (Witthoft and Farver 1952) and the Stewart site in Clinton County (Witthoft 1954). Kinsey and Graybill's (1971) review comprises the most recent overview of the complex from a regional perspective. The traditional view of Shenks Ferry interprets the archaeological remains as indicative of a "culture" with three sequential phases: Blue Rock, Lancaster, and Funk (listed from oldest to youngest) dating to between circa A.D. 1100 and A.D. 1550 (Kinsey and Graybill 1971, 30–36; Heisey 1971, 46, 58, 62–63). The distribution of Shenks Ferry sites shown in figure 14 is based on Kinsey and Graybill's (1971, 5, fig. 1) original study and more recent data (Custer, McNamara, and Ward n.d.). Within the area noted in figure 14, similarities in ceramic design motifs are quite marked, especially in the case of Shenks Ferry cord-marked and Shenks Ferry incised ceramics of the early Blue Rock phase (Heisey 1971, 65–66). On the basis of these marked similarities in ceramics, numerous scenarios of migrating, expanding, and contracting populations have been advanced (Heisey 1971, 64–68; Clark 1980). However, a more careful analysis of Shenks Ferry adaptations and community patterning reveals a more complex situation. These aspects of Shenks Ferry societies are considered here; however, Shenks Ferry interaction networks as revealed through ceramic distributions will be considered first.

Ceramics and Interaction Networks

Heisey (1971) has provided the most current description of Shenks Ferry ceramics and notes the basic design motifs and their decorative combinations. During the initial Blue Rock phase, when Shenks

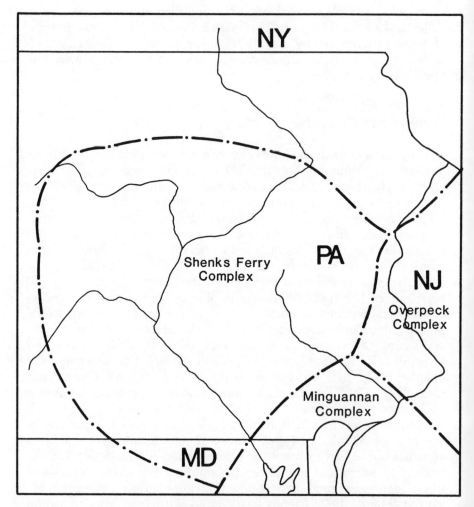

Fig. 14. Shenks Ferry complex distribution.

Ferry ceramics show the greatest similarities throughout their distribution, Shenks Ferry ceramics are found in association with numerous Owasco and Clemson Island ceramics. Bressler (1980, 55) notes this association at the Bull Run site and the Greys Run Rockshelter (Bressler 1975) in the West Branch Valley, and Hatch (1980) notes a series of transitional Clemson Island/Owasco/Shenks Ferry design motifs, as well as a cooccurrence of "pure" Clemson Island and Blue Rock phase Shenks Ferry ceramics in the stratified midden at the Fisher Farm site, also in the West Branch Valley. A similar cooccurrence of Owasco ceramic varieties and Shenks Ferry ceramics is seen in the North Branch Valley (Lucy 1959). The Nace site (Holzinger 1970), a section of the Blue Rock site (Heisey and Witmer 1964) in the Lower Susquehanna Valley, also produced a ceramic assemblage of primarily Shenks Ferry incised and Shenks Ferry cord-marked varieties associated with Owasco corded, Bainbridge incised, Levanna cord-on-cord, and Kelso corded ceramics. Thus, at least during the early phases of the Shenks Ferry complex, interaction with groups to the north and west continued. The appearance of the later Owasco-related varieties without Clemson Island varieties at the Nace site in the Lower Valley suggests that by circa A.D. 1200 Lower Valley groups were also drawn into the same interaction networks as groups in the Middle Valley. However, a consideration of Shenks Ferry ceramic design motifs, vessel shapes, and rim treatments show that other interregional networks of interaction were even more important during the initial Shenks Ferry phases.

The previously mentioned transitional Owasco/Shenks Ferry ceramics from Fisher Farm (Hatch 1980, 301–05) here indicate the combining of new design motifs with older forms. Although this combining of motifs may indicate that Shenks Ferry motifs are derived from Owasco motifs, a cursory analysis of these two sets of motifs suggests that this is not the case (compare Heisey 1971, 53—fig. 28a-h with Hatch 1980, 284—fig. 18.4; 268—fig. 18.5; 287—fig. 18.6; 290—fig 18.8; and also Turnbaugh 1977, 235). In general, Shenks Ferry ceramics are more similar to ceramics from areas to the south and southwest. In the original description of Shenks Ferry cord-marked and Shenks Ferry incised ceramics, Witthoft and Farver (1952) note that these wares are similar to Page and Shepard ceramics of the Piedmont of western Maryland and northwestern Virginia. Similarities to Monongahela ceramics have also been noted by Kent (1974, 5). Incised varieties have also been viewed as similar to Fuert Focus, Fort Ancient ceramics (Witthoft and Farver 1952; Griffin 1943, 193–94, 343–47). R. M. Stewart (1982a), in summarizing ceramics in Maryland's Hagerstown Valley, notes the relationships among Shepard, Page and Keyser ceramics and the similarities to both Shenks Ferry and Mononghela wares. Finally, ceramics from the Monocacy Valley in the Maryland Piedmont (Peck 1979; Peck and Bastian 1977) summarized by Kavanagh (1981) fit within a stylistic continuum spreading from the Great Valley into the Piedmont of Maryland and Pennsylvania. The poorly defined Montgomery Focus (Schmitt 1952, 62) also contains ceramics similar to the Shenks Ferry cord-marked series as noted by Clark (1980, 16).

Thus, although groups inhabiting the Middle and Lower Susquehanna Valley continued to interact with groups to the north, the primary emphasis in interaction networks shifted to the south and west. Nevertheless, in spite of the interaction of groups within the central Middle Atlantic Piedmont and Great Valley, the Middle and Lower Susquehanna Valley emerged as an area of intensive local interaction that produced the distinctive Shenks Ferry archaeological complex. At the same time (the Blue Rock phase), significant ceramic and cultural differences also emerge between the Lower Susquehanna Valley and the Minguannan complex societies of the Piedmont Uplands and High Coastal Plain (Custer, McNamara, and Ward n.d.). Early Shenks Ferry groups seem to have seldom interacted with groups to the southeast.

The patterns of interaction described above have implication for some migration theories of Shenks Ferry origins. For example, Clark (1980, 16) has suggested that Shenks Ferry groups were an intrusive population derived from the Virginia and Maryland Piedmont. However, the extensive analyses of Shenks Ferry ceramics by Witthoft, Stewart's more recent review, and data from the Monocacy Valley (Kavanagh 1981) reveal more complex ceramic associations and distributions than Clark describes. This complexity precludes the identification of "cultures" and migrations based on ceramic design distributions. Also, the chronology of the distributions (R. M. Stewart 1982a; Heisey 1971) do not match the basic assumptions underlying Clark's hypotheses. Similar problems are also apparent in Heisey's (1971, 64–68) scenario of cultural developments (Bressler 1980, 62). In sum, the Shenks Ferry complex is a local development within the Lower and Middle Susquehanna Valley. An intensive interaction network, indicated by similar ceramic design motifs, emerged in the Lower and Middle Susquehanna Valley. At the same time, regional interaction networks shifted from the north and west to the south and west. Nevertheless, during early stages (up to at least circa A.D. 1250) some interaction with groups to the north continued.

During the Lancaster and Funk phases, another shift in interregional interaction networks occurred. The local networks of interaction were neither tightly linked nor extensive, as indicated by the virtual absence of Lancaster and Funk incised ceramics in the Middle Susquehanna Valley and in the North and West Branch Valleys (Turnbaugh 1977, 235; Heisey 1971, 65–66; Kent 1980, 103). Although the "Wyoming Valley Culture" (I. F. Smith 1973) can be hypothesized as a replacement for Shenks Ferry groups in the North Branch Valley, no archaeological complexes dating from the time period between late Blue Rock phase times and the appearance of the Susquehannock in the middle 16th century are currently known in the West Branch Valley and the Middle Susquehanna Valley. Heisey (1971, 66) suggests that a southward population concentration may have occurred in post-Blue Rock phase times; this hypothesis can be evaluated by a consideraton of Lancaster and Funk phase ceramics and other contemporary ceramic types of the central Middle Atlantic region.

Lancaster and Funk phase ceramics differ from Blue Rock ceramics by their incipient collars, notched lips, punctuates, and designs based on plats of triangular motifs (Heisey 1971, 58–63, figs. 31–34, table 3). Although the design motifs of Lancaster and Funk ceramics are similar to the basic design elements of Shenks Ferry incised, especially Shenks Ferry incised multiple-banded and complex varieties (Heisey 1971, 50, 53—fig. 28; 58, 64—fig. 35), the appearance of collars and the Funk incised designs show definite similarities to Susquehannock Schultz incised ceramics (Heisey 1971, 62; Kinsey 1959, 68–77, figs 6 and 7; Witthoft 1959, 42–51), and para-Iroquoian types such as Munsee incised, Ithaca linear, and McFate incised (Witthoft 1959, 50–51; Heisey 1971, 62; Guthe 1958, 59; Kent 1980, 99; I. Smith 1984). These similarities suggest that interaction networks of Shenks Ferry groups again extended to the north. Other social groups also appear to be drawn into interaction networks that include northern para-Iroquoian social units. In the Upper Delaware Valley (north of the Water Gap), Iroquoian ceramic types, locally defined as Munsee, as well as numerous Chance incised ceramics, are common after A.D. 1250–1300 (Kraft 1975b, 1978; Griffith and Custer 1983, table 4) and replace Overpeck and Bowmans Brook ceramics that show interaction networks extending to the south down the Delaware River Valley into the Coastal Plain (Staats 1977, 1974; Custer and Griffith 1983). In the North Branch Valley, Wyoming Valley ceramics, including Wyoming Valley incised, gashed, and lobed, and Parker incised (I. F. Smith 1973, 32–41), show marked similarities to Chance ceramics, as well as Munsee (Smith 1973, 51–54), and replace Shenks Ferry incised and cord-marked varieties along with some Northwestern Pennsylvania varieties such as McFate. In general, the overall tendency was for most societies of the Susquehanna Valley to be drawn into interaction networks including para-Iroquoian societies at the expense of the relationships to the southwest. However, Kent's 1974 analysis of Locust Grove ceramics suggests that some interaction with groups to the west continued through the Lancaster and Funk phases. Figures 15 and 16 summarize the changing interregional interaction patterns.

Subsistence Systems, Settlement Patterns, and Community Patterning

The traditional view of Shenks Ferry settlement/subsistence systems stresses the role of agriculture and settled villages (Kinsey and Graybill 1971, 35–37; Heisey 1971, 53; Turnbaugh 1977, 236). How-

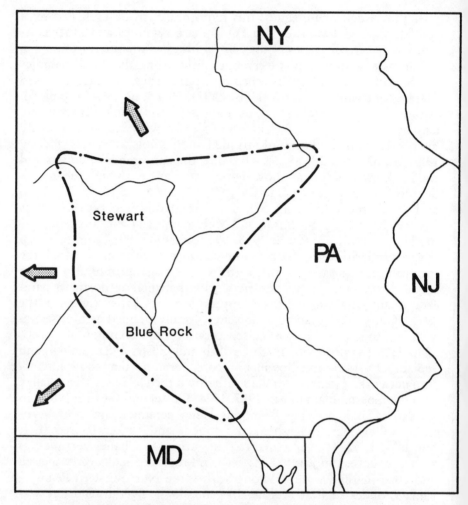

Fig. 15. Blue Rock/Stewart phase distributions and interaction.

ever, Graybill's more recent analysis of Shenks Ferry settlement pattern data suggests that the Shenks Ferry sites in the Lower Susquehanna Valley were of two general types:

> ... year-round villages and seasonal encampments. Both types were occupied during Lancaster-Funk times, with seasonal camps characteristic of the Blue Rock Phase. ... there is presently no evidence to indicate that any Blue Rock site was occupied on a year-round basis. Rather, the recurrent-seasonal nature of Blue Rock sites is suggested by the seasonal character of food remains, ceramic diversity, and burial patterns. (1973, 8)

Graybill (1973, 12–16) goes on to describe a series of excavated sites that support this contention and also considers surface survey data noted in his earlier study that was coauthored with Kinsey (1971, 30–35). In general, the seasonal encampments occupy a variety of topographic settings including both sheltered and open locales and appear as randomly dispersed refuse patterns with restricted inventories of tool types (Graybill 1973, 12). On the other hand, the later villages

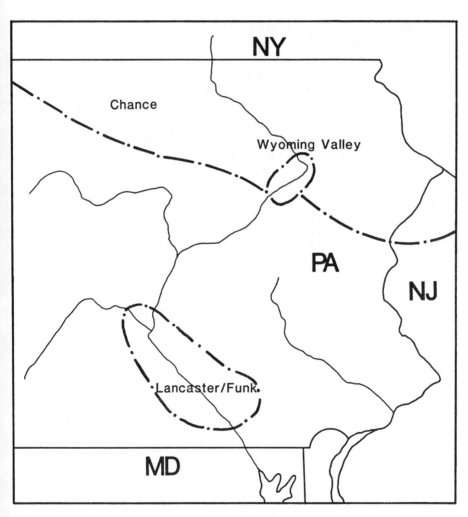

Fig. 16. Lancaster/Funk phase distribution and interaction.

... occupy well-drained settings in proximity to soils which are high in natural fertility, typically along streams paralleled by alluvial bottomlands; favor southeastern exposures; exceed two acres in areal extent; appear as donut-shaped configurations of refuse (especially the latest villages); and exhibit tool types and food remains indicative of multiple subsistence activities (1973, 12).

It is especially interesting to note that the Blue Rock phase settlement pattern described by Graybill for the Lower Susquchanna Valley shows a similarity to Early and Middle Woodland patterns of the Lower Susquehanna Valley region and the Late Archaic–Late Woodland settlement patterns of the adjacent Piedmont Uplands area (Custer and Wallace 1982). In the Piedmont Uplands area of northern Delaware and southeastern Pennsylvania there is good evidence for the *absence* of agriculture and sedentary villages (Custer and Wallace 1982; Custer 1982a, n.d.) throughout the Late Woodland period; moreover, hunting and gathering lifeways persist in the area up through the time of European Contact (Becker 1980a). A similar pattern seems to apply at least to the pre-A.D. 1300 period in the Lower Susquehanna Valley. Even in the excavated rockshelters of this region, where there is good organic preservation (Kurtzman 1974; Kent and Packard 1969), there are few, if any, indications of any kinds of cultigens in Late Woodland contexts.

As a test of Graybill's summary of Shenks Ferry settlement patterns, I undertook a reanalysis of Late Woodland site locations in the northwestern Piedmont Uplands and Lancaster Valley. The site files of the Section of Archaeology, William Penn Memorial Museum, were utilized as a data base, along with the sites noted in the Eckman Collection (Kinsey 1977, 381), housed at the North Museum, Franklin and Marshall College. A total of 215 sites with identifiable components were used for analysis. Although the data base is biased, statistical analysis of general trends in the data can be carried out. However, formal inferential statistics are not applicable. Table 5 summarizes the cross-tabulation of the sites by time period of occupation and associated surface water of various orders. The time period categories included in table 5 are sites with pre-Late Archaic components, multicomponent sites with Late Archaic–Late Woodland components, and sites with only Late Woodland components. These categories were chosen to see if there were any significant differences between Late Archaic–Middle Woodland site locations and Late Woodland site locations. Following Graybill's description of Shenks Ferry settlement patterns, one can hypothesize that there should be minimal differences between Late Archaic–Middle Woodland and

Table 5
Lancaster Lowland Site Data

Surface Water Setting	Pre-Late Archaic	Late Archaic–Late Woodland	Late Woodland
High Order	10	14	13
Medium Order	33	14	19
Low Order	40	17	7
Interior (no surface water)	30	12	6

Late Woodland site distributions. More specifically, if there were a shift to higher-order drainage floodplain areas, it should be apparent, as is the case for other major river valleys of the Middle Atlantic (Gardner 1982, 78–80; R. M. Stewart 1982b).

In order to evaluate these hypotheses, a chi-square analysis was carried out, using the data from the Late Archaic–Late Woodland multicomponent column and the single component Late Woodland column. If there were a shift in settlement patterns associated with agriculture, then the chi-square test should show significant differences indicating concomitant variation between the variables of time period and associated surface water. The chi-square statistic equals 5.02 with 6 degress of freedom ($.50>p>.25$). Therefore, neither significant difference nor concomittant variation is indicated. Consequently, the overall site data from the Lancaster Valley do not contract Graybill's contention that during early Late Woodland times (up through Blue Rock phase) agriculture did not play an important subsistence role and that site distributions were very much like those of earlier periods. Closer analysis of the thirteen single-component Late Woodland sites located in high-order drainage floodplains shows that all are either Susquehannock or Lancaster-Funk phase Shenks Ferry sites. This late Late Woodland association of sites with high-order drainage floodplains also supports Graybill's hypothesis.

Few seasonal camp and/or procurement sites from the Blue Rock phase have been excavated; however, some insights into community patterning can be developed by considering those available for study. Graybill (1973, 15) notes four major Blue Rock phase sites from the Pequea drainage, three of which are small rockshelters. The only open site noted has not yet been extensively tested. In general, the rockshelter sites are quite small (see Kent and Packard 1969) and were probably inhabited by no more than a nuclear family. An additional habitation site of the Blue Rock phase is the Blue Rock/Nace site complex (Heisey and Witmer 1964; Holtzinger 1970). At the Blue Rock site clusters of refuse/storage pits, hearths, midden areas, and graves were discovered and clearly dated on the basis of ceramics to the Blue Rock phase (Heisey and Witmer 1964, 11–15). No postmolds or indications of permanent houses were present, and the excavators considered the clusters of pits to indicate the presence of small residential areas that may, or may not, have been occupied contemporaneously (Heisey and Witmer 1964, 32–34).

Excavations at the adjacent Nace site by Holtzinger (1970) provide a clear view of intrasite settlement patterns. In an area approximately 25 feet by 25 feet, one and possibly two houses approximately 18 feet in diameter were defined on the basis of postmolds and shallow organic-stained depressions. Associated with each house were graves, hearths, amd small sheet middens, some of which were as much as 32–36 inches thick. Holzinger (personal communication, March 1983) feels that the two sets of features were not occupied at the same time and, indeed, notes that their shapes are quite different. One appears to be circular, while the other resembles the "keyhole" structures of the Middle and Upper Susquehanna Valley (I. F. Smith 1976; Hatch and Daugirda 1980). My interpretation of the Nace data is that it represents two noncontemporaneous residential units with associated food storage, preparation, disposal, and burial activity areas. Each residential unit would correspond to a "household cluster" as defined in K. V. Flannery (1976) and probably was occupied by a single nuclear family, or at most a pair of nuclear families. The presence of the entire range of domestic processing, storage, and disposal areas at a single locus indicates that the nuclear family was the basic production and consumption unit. This view of the basic social organization of early Shenks Ferry societies in the Lower Susquehanna Valley corresponds to data from the rockshelter sites.

Data from the Sheep Rockshelter (Michels 1967; Green 1968) suggest that similar community patterns and organizations were present in the Juniata drainage. However, the Sheep Rockshelter floral data shows more use of cultigens (Steffy 1968). The Fisher Farm site in the West Branch Valley area shows a similar internal settlement pattern and has been described as a "hamlet" (Hatch and Stevenson 1980). However, as noted by Graybill (1982), the confusion generated by the multicomponent nature of the site may make this description questionable. Nevertheless, the site is clearly not a sedentary village and is similar to patterns noted by Graybill for the Lower Susquehanna Valley.

Archaeological data from the West Branch Valley proper provide a contrast to the previously described pattern. Bressler (1980, 36) notes two stockaded villages, the Bull Run site and the Wolf Run Fort, that contained Shenks Ferry cord-marked and incised ceramics that are part of the Stewart phase, a West Branch Valley Shenks Ferry phase contemporaneous with the Blue Rock phase (Witthoft 1954). Neither site was excavated to a sufficiently large extent to understand the entire community patterning; however, Bressler (1980, 37) estimates the size of the Bull Run stockade as an oval 200 feet long and 124 feet wide, enclosing an area of approximately 1.7 acres. In one case, the stockade was placed around an existing community. Flotation analysis of feature fill revealed some maize and a variety of wild plant foods including butternut, purslane, carpetweed, verbena, chickweed, oalis, and redmaids (Bressler 1980, 38).

To summarize to this point, during the earliest phases of the Shenks Ferry chronology, the Lower and Middle Susquehanna Valley was inhabited by a group of interacting societies who were living in dispersed, scattered hamlets with nuclear familes or limited sets of nuclear families, comprising the maximum units of food production and consumption. The association of burials with individual houses, or household clusters, also indicates that certain rituals were carried out by the minimal social units and were not managed and organized on a supralocal community basis. The subsistence base of these groups was primarily hunted and gathered resources; however, in the West Branch Valley and other areas to the north and west, some cultigens were present. The stockaded villages in the West Branch Valley, coupled with the use of cultigens, indicated that more complex organizations occurred in this area. Fortification of existing communities comprised of multiple social units seems to indicate that larger communities than those of the Middle and Lower Valley existed prior to a need for fortification. The presence of these larger, partially agricultural, semisedentary communities is understandable in light of the existence of the previously described earlier Owasco/Clemson Island villages and late Middle Woodland semisedentary communities of areas to the north and west. The continued interaction with these other areas, as indicated by ceramic styles, would have brought cultigens into the area, which would have allowed the development of larger, more sedentary communities. The addition of stockades may have been in response to intergroup competition generated by the appearance of higher population densities in the agriculturally productive floodplain settings. These differences in community patterning and subsistence activities noted between the societies using Shenks Ferry ceramics during Blue Rock and Stewart phase times in the West Branch Valley and Lower and Middle Susquehanna Valley highlight the difficulties inherent in defining *archaeological cultures* in terms of ceramics and seriously call into question the notion of a "Shenks Ferry Culture."

Beginning in Lancaster phase times (after circa 1250 A.D.), sedentary villages appeared in the floodplains of the major drainages in the Lower Susquehanna Valley. Graybill (1973, 12–15) notes seven village sites in the Pequea drainage and Kinsey and Graybill (1971, 31, fig. 21) note several others on the main branch of the Susquehanna and in the Conestoga drainage. Small seasonal encampments still continued to be used and probably represent specialized procurement sites. An especially good example of such sites is the Upper Bare Island Rockshelter, which has been identified as a fishing camp based on the presence of sturgeon remains (Graybill 1973, 15).

Several village sites from Lancaster-Funk phase times have been subjected to various levels of excavation and reveal some intrasite settlement patterns and subsistence data. Excavations at site 36LA240 at the confluence of the Conestoga and Cocalico creeks by Elvin Geltz of Lancaster revealed an extensive midden partially deposited in a sinkhole. Although a thorough, professional analysis of the materials has never been allowed, maize remains from the site have been observed by this author and are described as extensive (Elvin Geltz, personal communication, July 1979). No data are available on any associated habitation areas and the property owner will not allow access for further survey work. Excavations at the Kibler-Funk site in Washington Boro (Snyder 1975) showed some storage and refuse features associated with a plow-disturbed living area. An extensive subsurface midden associated with the site was tested; however, since the organic preservation was poor, no ecofacts turned up.

The most extensive excavations of a Lancaster-Funk phase village were carried out at the Murry site, reported by Kinsey and Graybill (1971). The Murry site is a stockaded village encompassing almost four acres and having a radiocarbon date of A.D. 1410. Over 60 percent of the village was excavated, and fifty-two houses arranged in two concentric circles were projected (Kinsey and Graybill 1971, 23–28). The houses are oval and approximately 20–25 feet long and 13–15 feet wide, making the average area within each house 320 square feet. Kinsey and Graybill (1971, 30) estimate the population of the village at approximately 550 people. Houses usually have associated storage/refuse features and hearths, and burials within houses are common (Kinsey and Graybill 1971, 27—fig. 17). No extensive midden remains turned up within the village and testing outside the village was limited. Therefore, no midden-derived subsistence data are available. A large circular structure at the center of the village with no associated features and few artifacts may be a men's lodge or ceremonial structure (Kinsey and Graybill 1971, 28–29) and may indicate a focal point for community organization. The Locust Grove site (Kinsey and Graybill 1971, 37–38; Kent 1974) and the Mohr site (Gruber 1969, 1971) are partially excavated, stockaded villages of the Lancaster-Funk phase similar to Murry.

The household cluster of houses, storage/refuse features, hearths, and burials within houses seen at the Murry site are similar to the patterns of the Nace site. Thus, the Murry site seems to represent an amalgamation of the nuclear family production/consumption units of earlier times. The "men's lodge/ceremonial structure" at the center of the village and the absence of individual refuse middens suggest the beginning of a suprafamily community organization. More important, the integrated village plan and stockade at Murry shows detailed community planning. However, the burials in individual houses suggest that community integration did not extend to ritual events such as interment.

Analyses of burials from the various Shenks Ferry sites indicate that Shenks Ferry societies were essentially egalitarian. Grave goods are uncommon and few in number, and when present are primarily found with adult males and children (Kinsey and Graybill 1971, 13–18; Heisey and Witmer 1964, 15–19; Witthoft and Farver 1952; Gruber 1971), indicating that for special roles, individuals were recruited by age and sex. Status was most likely achieved rather than ascribed. Gruber (1971) and Kinsey and Graybill (1971, 17–18) note that orientation of graves is primarily to the east—a pattern that extends through all known sites with burial complexes in the Lower Susquehanna Valley. Kinsey and Graybill (1971, 29) also note that the opening to the ceremonial structure at Murry also faces the east, and they suggest an ascribed importance to this direction in Shenks Ferry cosmology. It is interesting to note that this pattern includes Shenks Ferry sites of all phases, which suggests a cosmological continuity through time. However, it is not clear that this pattern extends throughout the entire distribution of Shenks Ferry sites beyond the Lower Susquehanna Valley.

In sum, even with the appearance of sedentary villages, the basic underlying social organization of Shenks Ferry groups does not change appreciably from an egalitarian nuclear family base. Suprafamily organizations sufficiently permanent enough to plan and construct stockaded villages like Murry certainly existed, but these organizations apparently did not affect food production and consumption or rituals such as burial of the dead. All finds of cultigens from Lancaster-Funk phase sites are not from the best context, but it is highly probable that cultigens played some role in subsistence. The stockades suggest that some intercommunity hostilities existed, which may have been the result of competition for scarce resources in the face of rising population densities. However, by the middle of the fifteenth century, apparent population dislocations may also have induced hostilities. These trends are best considered in relation to the Susquehannocks.

Susquehannock Complex

The Susquehannocks are probably the best-studied aboriginal society in the Middle Atlantic from both an archaeological and historical perspective (Kent 1984). A fairly detailed ethnohistoric data base (see Jennings 1978, 367) and the linking of ethnohistoric site descriptions with archaeological sites (Witthoft 1959; Hunter 1959) allow the application of the direct historical approach that traces the migration of the Susquehannocks down the Susquehanna Valley from southeastern New York (Witthoft 1959; Kent 1980; Jennings 1978). Barry Kent, Pennsylvania's State Archaeologist, has completed a book on the Susquehannocks and in deference to this wider and more complete study (1984), I will here discuss only three topics: the relationships between the Shenks Ferry and Susquehannock complexes, Susquehannock subsistence systems, and Susquehannock community patterns and social complexity.

Shenks Ferry and Susquehannock Relationships

Note that the Susquehannock archaeological complex can be clearly linked to a distinctive cultural entity. Using ethnohistoric data, one can define Susquehannock community patterns, subsistence systems, social organization, ideology, material culture, and language (Jennings 1978), which are clearly derived from an Iroquoian base. It is also generally agreed that this culture did not develop in the Lower Susquehanna Valley and that the Susquehannocks moved into this area during the second half of the sixteenth century (Jennings 1978, 362; Kent 1980, 99). However, reasons are cloudy for the movement, its relationship to other population disruptions, and the fate of the indigenous Shenks Ferry groups.

Kinsey and Graybill (1971, 39) suggest that the intrusion of Susquehannock groups into the Lower Susquehanna Valley may have forced the coalescence of social groups that produced the stockaded Shenks Ferry villages seen during Lancaster-Funk phase times. Based on data from the Murry site, including finds of a "trophy skull," a mass burial, and indications that at least part of the village may have been destroyed by fire, they also suggest that the inhabitants of the Murry site may have been at war with the Susquehannocks. However, while there can be no doubt that conflicts took place during Lancaster-Funk phase times, the radiocarbon date of A.D. 1410 ± 100 years from the Murry site (Kinsey and Graybill 1971, 39) is almost 140 years before the first archaeologically recognized appearance of a Susquehannock cultural entity in the Northern Branch Valley (Kent 1980, 99; Witthoft 1959, 32). Therefore, although the Susquehannocks eventually displaced and/or destroyed the indigenous cultures of the Susquehanna Valley, this destruction was not the major cause of hostilities during the later phases of the Shenks Ferry complex.

The movement of the Susquehannocks can be understood as part of a pattern of population disruptions that seems to have begun during the fifteenth century. As noted earlier, during the Lancaster-Funk phase times, Shenks Ferry groups of the Lower and Middle Susquehanna Valley intensified their interactions with northern groups. This intensification had begun earlier in the West Branch Valley and had involved warfare. Similar patterns also characterized the Upper Delaware River Valley, although there are no stockaded villages in this area. In the North Branch Valley, groups of the Wyoming Valley complex were also drawn closely into interaction networks with northern Iroquoian groups by at least A.D. 1475 (I. F. Smith 1973, 41), and well-developed fortifications are present (e.g., the Scacht site—I. F. Smith 1973, 46–47), indicating that conflict was also present or threatened. Similar developments occurred in the West Branch Valley and in northwestern Pennsylvania (I. Smith 1984; Johnson, Richardson, and Bohnert 1979). In general, there seems to be a correlation of stockaded villages, warfare, agriculture, and interaction with northern Iroquoian groups throughout the upper reaches of the Susquehanna Valley. Although the conflict may be among local groups exclusive of Iroquoian societies, the conflict seems to intensify with increased northern interaction during the last two centuries prior to European Contact and extends well into the Lower Susquehanna Valley.

It is interesting to note that these conflicts and other events can be correlated with the onset of the "Little Ice Age," a moist and cool climate perturbation (Carbone 1982, 47, fig. 3). Griffin (1961) notes the correlation of climatic change, cultural conflicts, and population disruptions in the circum-Great Lakes area, and I hypothesize here a similar relationship for the Susquehanna Valley. Especially in the northern areas of the drainage, agricultural productivity was limited by the number of frost-free days (see Sykes 1980). If the "Little Ice

Age" induced lower temperatures that significantly reduced the number of frost-free days, as it did in other areas in comparable latitudes (Bernabo 1981), reductions in crop yields, especially maize, would have been likely (Griffin 1961). These disruptions of subsistence systems may have increased the potential for conflict over scarce resources, such as arable land, throughout the Susquehanna Valley and in other areas of the Middle Atlantic as well (Clark 1976).

Thus, the movement of the Susquehannocks south into the Lower Susquehanna Valley may have been the result of the hostilities endemic to the fifteenth and sixteenth centuries induced by climatic change in the central Middle Atlantic. Witthoft suggested some time ago (1959, 35) that the Susquehannocks moved south due to their declining political situation in the north or possibly because of a military defeat at the hands of their Iroquoian neighbors, and I support this contention here. However, Hunter (1959, 13) notes an additional factor that may have drawn Susquehannock groups south. By the middle of the sixteenth century, European goods were finding their way into the hands of Iroquoian groups (Wray and Schoff 1953). The marginal location of the Susquehannocks with regard to the early Saint Lawrence River fur trade placed them at a disadvantage to their neighbors, and indeed, this disadvantage may have been the cause of their declining fortunes. Given the fact that European contact had begun in the Chesapeake Bay area by as early as 1568–80 (Feest 1978b, 254, fig. 1), the Susquehannocks may have been drawn south by the potential trade opportunities (Hunter 1959, 13). The final choice of the Washington Boro area was probably due to the highly productive soil and particularly long growing season of the area documented by Parry (1975).

Kent (1980, 99) notes that the movement down the Susquehanna Valley took place over a fifteen-year period (1560–75). Prior to 1560 the Susquehannocks were living in dispersed hamlets in the Upper Susquehanna Valley, but by 1575, a date determined by European trade goods, they had coalesced into a single stockaded town, the Schultz site, in Washington Boro, Lancaster County (Kent 1980, 99). At the same time,

> ... there is complete termination of all other native cultures in the Susquehanna Valley; no other groups anywhere along the river in Pennsylvania survived long enough to have acquired trade goods. The Susquehannocks were primarily responsible for dispersal or termination of protohistoric Wyoming Valley folk, those of the West Branch, and Shenks Ferry in the Lower Valley. (Kent 1980, 103)

Although the disappearance of the Shenks Ferry groups is obvious from the archaeological record, the mechanism of their disappearance is not clear. However, some processes and events apparent from the archaeological record do provide some insights.

Over the fifteen-year period of their movement from the Upper Valley to the Lower Valley, the Susquehannocks changed their settlement pattern from one of a series of dispersed hamlets to a single fortified town, the Schultz site (Kent 1980, 99). At present there is insufficient information to develop direct population estimates for the Susquehannocks, prior to 1560; however, using indirect data, we can develop some projections of general trends in population growth. A population sufficiently large to allow the conquest or dispersal of the Wyoming Valley societies had to have existed in the initial Susquehannock societies that began the migration down the Susquehanna Valley. Given the fact that the Wyoming Valley villages were at least as large as Shenks Ferry villages and had similar internal settlement patterns (compare I. F. Smith 1973, 10—fig. 5, 46—fig. 30 with Kinsey and Graybill 1971, 28—fig. 19), and given a population estimate of at least 550 for the Murry site (Kinsey and Graybill 1971, 30), the initial Susquehannock population must have been at least as large. The absence of early Susquehannock villages suggests that the population was not too much larger; however, by 1580, when they settled at the fortified Schultz site, their population is estimated at between 1,000 and 1,500 (Casselberry 1971, 180). Based on this rather tentative analysis, a significant growth in population is hypothesized for the fifteen-year period of the Susquehannock migration. This significant growth may indicate that local indigenous populations were incorporated into Susquehannock communities as the migration progressed, creating a "snowball effect" of population growth that culminated in the large community at the Schultz site. Such a pattern of population growth would indicate that possibly some indigenous Shenks Ferry people were incorporated into the Susquehannock society. Witthoft (1959, 23–25) suggests a similar interaction, based on ceramic design data, and Kent (1980, 99) also notes that interaction with other southern and western groups is evident in the earliest Schultz incised ceramics found at the Schultz site. Although Kinsey and Graybill's 1971 Murry site data suggest that the extent of interaction was not as great as originally hypothesized by Witthoft, the interaction and incorporation of social groups was probably a significant feature of Susquehannock society. Therefore, in spite of the fact that the coming of the Susquehannocks brought about the demise of the indigenous societies as distinctive social units, some members of indigenous groups of the Lower and Middle Susquehanna Valley probably contributed to Susquehannock culture.

Subsistence Systems

Most investigations of Susquehannock sites have focused on living areas within village stockades and cemeteries and have not generated much subsistence data. However, some data are available from excavations in the midden associated with the Washington Boro village, dated between 1600 and 1645 (Kent 1980, 103); from informal studies of the Strickler site midden, which is dated between 1645 and 1665 (Kent 1980, 103); and from analysis of food remains preserved in brass kettles from the Strickler site that were exposed and investigated when the Pennsylvania Department of Transportation crews cut into the midden while widening River Road in Washington Boro. Arthur Futer, an avocational archaeologist from New Holland, was in charge of the highway crews and collected materials from the exposed midden before it was destroyed. These materials, which were then donated to the North Museum, Franklin and Marshall College, include various artifacts such as large storage vessels and lithic debitage. More important, also recovered were a large number of charred corncobs. These finds also correspond to food remains preserved in brass kettles that were deposited as grave goods. Futer (1959, 139) notes that corncobs and seeds of pumpkins, squash, and beans were found. A reexamination of these materials, on deposit at the North Museum, reveals in many cases large shell fragments of pumpkins and squash with attached seed masses. Other preserved food remains included fish, tentatively identified as sucker (Catostomidae).

A more extensive analysis of the materials from the Washington Boro midden used ecofacts recovered from intensive excavations in the organic-rich midden. Guilday, Parmalee, and Tanner (1962) report an analysis of faunal remains from a section of the midden excavated by John Witthoft. Table 6 lists the wide variety of faunal species that were used as food sources. Also including an examination of butchering and skinning marks, the analysis shows that these marks often confirm the food utilization of certain species that seem intuitively unpalatable. Butchering marks also reveal the cuts of meat consumed. The largest animals (bear, deer, and elk) were butchered into units that weighed as much as thirty-five to fifty pounds (Guilday, Parmalee, and Tanner 1962, 68, 77, 79). These large cuts of meat were then processed by stewing or boiling. Witthoft (1959, 49) notes that some ceramic vessels have capacities of up to eighty-six quarts. In fact, a vessel from the Washington Boro village on display at the

Table 6
Faunal Food Sources from Washington Boro Midden

Mammals
gray wolf (Canis lupis)
gray fox (Urocyon cinderedargenteus)
black bear (Ursus americanus)
raccoon (Procyon lotor)
fisher (Martes penneti)
otter (Lutra canadensis)
striped skunk (Mephitis mephitis)
bobcat (Lynx rufus)
mountain lion (Felis concolor)
woodchuck (Marmota monax)
gray squirrel (Sciurus carolinensis)
fox squirrel (Sciurus niger)
muskrat (Ondatra zibethicus)
beaver (Castor canadensis)
cottontail rabbit (Sylvilagus, species?)
white-tailed deer (Odocoileus virginianus)
elk (Cervus canadensis)

Birds
common loon (Gavia immer)
pied-billed grebe (Podilymbus podiceps)
whistling swan (Olor columbianus)
trumpeter swan (Olor buccinator)
Canada goose (Branta canadensis)
blue/snow goose (Chen, species?)
mallard/black duck (Anas, species?)
redhead/scaup (Aytha, species?)
common goldeneye (Bucephala clangula)
bufflehead (Bucephala albeola)
merganser (Mergsu, species?)
ruffed grouse (Bonasa umbellus)
bobwhite (Colinus virginianus)
wild turkey (Meleagris gallopavo)
whimbrel (Numenius phaeopus)
passenger pigeon (Ectopistes migratorius)

Reptiles
snapping turtle (Chelydra serpentina)
map turtle (Graptemys geographica)
musk turtle (Sternotherus odoratus)
box turtle (Terrapene carolina)

Fish
long-nosed gar (Leposusteus osseus)
shad (Alosa sapidissima)
sucker (Catostomidae)
catfish (Ictalurus)
eel (Anguilla bostoniensis)
walleye (Stizostedoin vitrium)
sea bass (Roccus lineatus)

Amphibians
bullfrog (Rana catesbiana)
frog (Rana, species?)

SOURCE: Guilday, Paronales, and Tanner 1962, table 1.

North Museum has a rim diameter of 18.5 inches and is 18 inches high. Brass kettles are similarly large (Hampton 1977). All of these large vessels could have contained the large cuts of meat (Guilday, Parmalee, and Tanner 1962, 68).

Floral remains from another set of excavations by Fred Kinsey in the same midden have been analyzed and reported by Ameringer (1975). Table 7 lists the various plant foods present. This list is not as extensive as the variety of wild plant foods reported by Moeller (1975) for the Upper Delaware Valley; however, the variety of wild plants would supplement the cultigens noted previously. Wild plant remains preserved in brass kettles of the Heisey collection also show a similar range of species (Anderson 1973). In addition to floral remains, analysis of mollusk shells from the Kinsey excavations were also carried out (Hartnett 1973). Freshwater mussels (Elliptio errans) were found throughout the midden, extensively in some levels. Estimated meat weights based on shells from three-inch levels of a 5 foot square area averaged between five and ten pounds; however, one level produced shell yielding an estimated meat weight of almost one hundred pounds (Hartnett 1973, 5).

The available subsistence data show that Susquehannock groups were clearly agricultural and that the extent of use of cultigens seems to be greater than for the Shenks Ferry complex, even when differential site size and extent of excavation are taken into account. A wide variety of freshwater shellfish, hunted animals, and wild plant foods also supplemented these cultigens. The butchering patterns and the size of processing vessels suggest that communal food preparation took place.

Community Patterning, Social Organization, and Social Complexity

Ethnohistorical and archaeological data show that the basic social organization of the Susquehannocks was quite different from that of Shenks Ferry societies. Excavations at several sites (e.g., Kinsey 1957) indicate that the Susquehannocks lived in Iroquoian-style longhouses, some of which were up to 95 feet in length, although the mean length at the Schultz site is 77.5 feet (Casselberry 1971, 185). Based on Iroquoian models, these houses were inhabited by matrilineally related nuclear families (Trigger 1978) and may have held up to ten families within a single dwelling, according to Casselberry's (1971, 185) analysis of postmold patterns within structures at the Schultz site. Ethnographic evidence on Iroquoian groups indicates that the individuals inhabiting a single longhouse formed a cooperative labor group for food procurement, preparation, storage, and consumption (Trigger 1978, 59–63; Fenton 1978). Since the large cooking/storage vessels and large cuts of meat noted earlier would fit this pattern, I hypothesize a similar organization for Susquehannock societies. These Iroquoian households were also linked through matrilineages and matrilineal clans, which recruited communal labor parties for harvesting, planting, land clearing, trade expeditions, and warfare (Fenton 1978, 309–12).

If these Iroquoian-type organizations applied to the Susquehannock, as indeed they seem to (Jennings 1978), then the Susquehannocks would have enjoyed a distinct advantage over the indigenous inhabitants of the Susquehanna Valley. The matrilineal extended families and matrilineages would have provided large-scale organizations for cooperative labor ventures such as warfare that the fragmented nuclear-family organizations of the Shenks Ferry groups could not as easily provide. Several authors have also suggested that the matrilineal organizations developed due to the absence of males from villages for a part of the year while they participated in trading, hunting, and warfare (Trigger 1978, 60–61; Ember and Ember 1971). This type of organization would have also given the Susquehannocks a decided advantage for participation in the fur trade when compared to their non-Iroquoian neighbors such as the Lenape. These large-scale organizations would also have facilitated the management of subsistence systems and agricultural activities that supported populations of up to 3,000 at the later Susquehannock towns (Jennings 1978, 362).

Analysis of grave goods distributions from Susquehannock cemeteries (Cadzow 1936; Heisey and Witmer 1962; Futer 1959; Witthoft,

Table 7
Plant Food Sources from Washington Boro Midden

corn (Zea Mays)
pokeberry (Phytolacca americana)
raspberry (Rubus occidentalis)
grapevine (Ampelopsis arborea)
wild beet (Amaranthus hybridus)
pondweed (Potamogeton)

SOURCE: Ameringer 1975, table 1.

Kinsey, and Holzinger 1959; Kinsey 1960; Custer and Futer 1978) suggests that special status clearly was achieved with recruitment by age and sex. This organization would be purely egalitarian and the dominant status positions were held by males. Trigger (1978, 63) also notes that dominant male status positions balanced the female-oriented matrilineal social organization, which promoted "an equality based on the separation and complimentality" of sexual roles.

In conclusion, the prehistoric and historic Late Woodland societies of the Lower and Middle Susquehanna Valley were all organized at a tribal level of complexity, at most, in Service's (1962) sense of the term *tribal*, with perhaps during early Late Woodland times a more simple band-level organization. In all cases, organizations were egalitarian (Fried 1967). Social organizations tended to become more complex through time in conjunction with the increased use of agriculture and the increased incidence of warfare. The Lancaster-Funk phase communities like the Murry site probably represent the highest level achieved through the processes of local sociocultural evolutionary change. However, even this development of increasing relative social complexity may have been stimulated by hostile conflict with more complex Iroquoian groups to the north. In this case, the evolution of tribal organizations in the Lower and Middle Susquehanna Valley would clearly fit with Fried's (1975) scenario for the evolution of tribal organizations.

7
Late Woodland Cultural Diversity in the Middle Atlantic: An Evolutionary Perspective

JAY F. CUSTER

> ... of course, prehistoric archaeology is not very meaningful without some kind of cultural evolutionary theory (archaeologists, perhaps necessarily, seem to be somewhat immune, even in the United States, to the injunctions against evolution), and evolutionary classification itself becomes more real and meaningful as archaeology can make more and more suggestions as to the actual course of events in specific cases.
> —Elman Service, *Primitive Social Organization*

This final chapter summarizes the sequences of culture change noted during Late Woodland times in each of the areas discussed in earlier chapters. Areas not discussed directly in previous chapters are also considered. My goal is to describe a series of multilinear cultural evolutionary pathways (Steward 1955, 11–29) that can be compared and contrasted in a controlled manner (Eggan 1954). I hope that such comparisons will lead to a general understanding of how and why cultures changed during the Late Prehistoric period in the Middle Atlantic region. Also, I show that the study of the processes of culture change during late prehistoric times in the Middle Atlantic region has implications for the study of larger issues in anthropology, such as the origin and development of tribal and chiefdom levels of sociocultural complexity (Service 1962; Carniero 1981) and the origins of ranked societies (Fried 1967).

The summaries consider the ethnohistoric data for the varied areas of the Middle Atlantic and then categorize the inferred level of sociocultural complexity of the societies that existed in the region immediately prior to Contact. I also note the sociocultural complexity of late Middle Woodland groups in the same region. Finally, I discuss the Late Woodland archaeological data pertaining to the processes of culture change that transformed these groups from their Middle Woodland forms to those observed at Contact.

Levels of Sociocultural Complexity

The basic typology of sociocultural complexity used here is that of Service (1962), who describes four basic levels: bands, tribes, chiefdoms, and states. No examples of the most complex level—states—are present in the prehistoric Middle Atlantic; however, the other three levels seem to occur in the region during Late Woodland times. Use of Service's basic concepts and definitions, however, does not imply acceptance of his notions of how these organizations arise. For example, I will discuss the issue of whether tribal organizations are truly an evolutionary stage that develops autochthonously through amalgamation of bands (Service 1962, 111–20, 178–81), or whether they are a secondary development (Fried 1975, 99–105), in light of Late Woodland data from the Middle Atlantic.

Band-level organizations, as noted by Service (1962, 108–9, 1966, 7–8) and Steward (1955, 122–27), are characterized by hunting/gathering subsistence bases, relatively nonsedentary lifestyles, and small size. According to Service (1962, 108), their most significant characteristic is "that all of the functions of the culture are organized, practiced, or partaken of by no more than a few associated bands made up of related nuclear families." Economy is organized by these kinship-based units, as are other social and political activities. The nuclear family with unilineal, usually patrilineal, extensions pervades the social organizations. Steward (1955, 122–50) recognized two varieties of band-level organizations, the patrilineal band—renamed "patrilocal band" by Service (1962, 60)—and the composite band. However, Service (1962, 83–107) convincingly argues that composite bands, as defined by Steward, are usually the result of post-Contact population and social organization disruptions. Only two Late Woodland societies of the Middle Atlantic, the Lenape and the Munsee of the Delaware River Valley, were at a band level of organization during Late Woodland times and at Contact. The composite-band label may also be applied to these groups. Archaeologically, band-level societies can be recognized by an absence of large planned living sites, or villages.

Tribal-level organizations occur in much of the Middle Atlantic during Late Woodland times. Service (1962, 111–13) notes that tribes are associations of larger numbers of kinship units than bands; however, tribes are not merely a collection of bands. Tribes include a new set of integrative institutions beyond exogamy and marital residence patterns, which provide the main sources of societal integration among bands. Tribes generally contain pan-tribal sodalities, such as age grades, clans, lineages, and secret societies, which help to organize production and reinforce social solidarity (Sahlins 1968). Nonetheless, tribes still tend to be politically fragmented when compared to more complex levels of organizations (Service 1962, 323–24). Tribes are usually egalitarian (Fried 1967), are usually at least semisedentary, and have some kind of domesticated plants or animals. At the least they intensively utilize a limited range of subsistence resources, and their lifestyle is best described as "Neolithic" (Cohen 1977; Harris 1979, 85–92).

Within the tribal category are some distinctions of various levels of organizations. For example, Oberg (1955, 477–84) distinguishes between segmented and homogeneous tribes while Service (1962, 120–40) separates lineal and cognatic/composite tribes. The main distinctions are between those groups with unilocal postmarital residence, lineal descent reckoning, and pan-tribal sodalities and residence groups (lineages and clans) based on these rules, which are termed lineal or segmented tribes, and those lacking lineality within residential groups and/or sodalities, which are termed cognatic/composite or homogeneous tribes. Sahlins (1961) notes the competitive advantages of lineal and segmented forms, and some Middle Atlantic Late Woodland data indeed support this contention. In the archaeological record, cognative tribal organizations can be recognized by planned village communities and household clusters that seem to be limited to nuclear families or small extended families. Lineal tribes are identified archaeologically by the presence of dwellings or household clusters that seem to be associated with large extended families or lineages.

Chiefdoms represent the most complex social organizations of the Middle Atlantic region and are found in only a few places, such as the Powhatan Chiefdom of the Virginia Tidewater discussed by Turner in this volume. The concept of the chiefdom as an evolutionary stage, or level of sociocultural complexity, has been discussed by numerous authors including Steward (1948; Steward and Faron 1959), Service (1962, 170–77), Oberg (1955, 484–85), and Sahlins (1958). In a recent overview of the concept of chiefdoms, Carniero (1981, 45) suggests the following minimal definition: "A chiefdom is an autonomous political unit comprising a number of villages or communities under the permanent control of a paramount chief." Their distinctive attribute is the fact that the political organization of a chiefdom is supralocal and includes more than one community. Other attributes of chiefdoms include the presence of redistributive economies, development of integrative social mechanisms that are not solely a function of kinship, and well-developed ranked social organizations (Fried 1967).

There is considerable variation among the societies that are often cited as examples of chiefdoms (Service 1962, 152–54) and recognition of this variation has led to the development of definitions of different types of chiefdoms. Although several typologies of chiefdoms exist (see Steward and Faron 1959, 177; Renfrew 1974, 74; Milisauskas 1978, 165; V. Steponaitis 1978), the typology suggested by Carniero (1981, 47–48) is used here. Carniero's typology consists of minimal, typical, and maximal chiefdoms and was chosen because it seemed to highlight the variability among chiefdom level societies. In Carniero's typology, minimal chiefdoms are those societies that meet the minimal criterion of chiefdoms, but do not go far beyond them (Carniero 1981, 47). Typical chiefdoms exhibit many of the elaborations of redistributive economies and ranked social organizations, but are still clearly less than states in their complexity. Maximal chiefdoms are those large and complex societies that border on state-level organizations. For the Middle Atlantic region the minimal and typical types of chiefdoms are important. In the archaeological record, it is very difficult to differentiate between tribal- and chiefdom-level societies as noted by Turner in this volume.

Note, however, that the levels noted above are not viewed as inviolate, monotheistical types. As most people using these typologies remind us, the categories are really points on a continuum of variation of sociocultural complexity. However, for comparative purposes these terms provide fixed reference points of common meaning. It is in this sense that I use these terms in the following descriptions. When Middle Atlantic Late Woodland societies are compared from these fixed reference points, I hope some insights on their sociocultural evolution will become apparent.

Upper Delaware River Valley

As Kraft notes earlier in this volume, the ethnohistoric groups of the Upper Delaware River Valley, the Munsee, do not appear to have lived in true villages. Although the Munsee had a common language and at times multiple social units acted in concert for economic and political activities (Goddard 1978a, 216; Weslager 1972), we do not know if they were at a tribal level of organization when observed by Europeans during the seventeenth century. The presence of multiple-named social units that correspond to geographical locations (Goddard 1978a, 213–16, 235–39) is more typical of band organizations. However, group sizes are somewhat larger than those of patrilocal bands and the Munsee clearly had a matrilineal kinship system, which is not at all typical of bands. Therefore, it is possible that the ethnographic Munsee may have been at a composite band level of organization and are not simple patrilocal bands.

The archaeological record for the Late Woodland period in the Upper Delaware River Valley argues with the ethnohistoric observations. No large villages are known from the area and the dominant settlement pattern seems to be a series of dispersed farmsteads. Domesticated plant remains indicate some reliance on agriculture; however, use of wild plant foods and hunted resources also seems to be an important component of the subsistence system (Moeller 1975). Extensive use of storage facilities would mean some degree of sedentism within the floodplains of the major drainages at the farmstead communities. However, where data on seasonality are available (Moeller 1975), there seems to be no year-round occupation of the sites. In sum, the archaeological data from the Upper Delaware River Valley show a relatively stable subtribal level of organization for most of the Late Woodland period.

However, throughout this period ceramic design data (Kraft 1975b; Puniello 1980; Griffith and Custer 1983) show that the societies of the Upper Delaware Valley were drawn increasingly closer into the interaction networks of more complex para-Iroquoian groups to the north and northwest. These changes in interaction networks may have had significant effects on regional sociopolitical organizations and may help to explain some possible anomalies of social organizations in the Upper Delaware, such as the presence of matrilineages. Service (1962, 83–107) suggests that during interactions between band-level organizations and more complex organizations, the resultant weakening of lineal kinship ties will result in the formation of composite bands. If this hypothesis is applied to the Upper Delaware, it is possible that interactions with para-Iroquoian groups caused the breakdown of some kind of unilineal band organization over the course of the Late Woodland period in the Upper Delaware. However, matrilineages and limited horticulture are not typical characteristics of bands, be they lineal or composite. These attributes of Upper Delaware groups may be related to a late influx of linked cultural innovations, such as matrilineality and incipient maize agriculture (Witthoft 1949) that were superimposed over the composite band adaptations. The archaeological data presented by Kraft (1975b) do suggest a late adoption of agriculture, as does the absence of distinctive settlement pattern shifts that are usually associated with the more intensive forms of agriculture.

Another possibility is that at one time the Late Woodland societies of the Upper Delaware did have an incipient tribal organization with matrilineages. Or, as Martin (1974) notes, it may be possible that matrilineal organizations developed along with intensified subsistence systems including gathering and fishing during Middle Woodland times and that the matrilineal organizations were associated with band-level organizations. In this scenario it is possible that if tribal level organizations were present, later interactions with the more complex Iroquoian groups caused these organizations to revert back to composite band levels following the interaction patterns noted by Service (1962, 83–107). Nonetheless, matrilineal organizations persisted. For the present, the two options should be viewed as equally competing hypotheses. Nevertheless, there is a general trend of social change due to interaction with more complex groups, along with a relatively stable adaptation at a subtribal level of organization in the Upper Delaware during Late Woodland times.

Middle and Lower Delaware River Valley and Upper Delmarva Peninsula

The ethnohistoric data from this portion of the Middle Atlantic (Weslager 1972; Goddard 1978a) and the data presented in this book by Becker suggest rather strongly that the Lenape had a band-level organization at the time of European Contact. Nevertheless, some authors (Thurman 1974) have noted that land sales, treaties, and communal hunts were carried out by multiple band-sized units, and that many of the Lenape clans could be viewed as "pan-tribal" sodalities, the existence of which Service viewed as an important attribute of tribal-level organizations (1962, 113). However, these descriptions, which suggest something more complex than a band-level organization, generally postdate 1680 and describe the Lenape after at least fifty years of European Contact and after perhaps as much as one hundred years of interaction with the more complex Susquehannocks (Jennings 1966, 1968). Thus, these later descriptions may not have much validity for the pre-Contact Lenape and do not refute the inferences of band-level organizations suggested here for them.

The Late Woodland archaeology of this area is quite varied in its implications for social organizations, but in general it is congruent with the ethnohistorical data. As noted by Stewart, Hummer, and Custer in this volume, the archaeological data from the Upper Delmarva Peninsula and the Lower Delaware River Valley show an absence of agriculture, very little use of storage, and an absence of sedentism. If anything, there is a decrease in social complexity in the

Upper Delmarva Peninsula and Lower Delaware River Valley during late Middle Woodland and early Late Woodland times. In contrast, Late Woodland sites in the Fall Line zone of the Delaware River Valley reveal that villages may have been present in this area. Also, there is a distinct settlement pattern shift that might show that agriculture played some role in the subsistence systems of Late Woodland groups in the Fall Line area. In light of these special features, it is likely that the Late Woodland societies in this area were at a tribal level of organization. The Middle Delaware River Valley, north of the Fall Line, contains sites that are similar to those of the Upper Delaware Valley area. However, preliminary data seem to show that agriculture was added to subsystems of groups living in the Middle Delaware River Valley even later than in the case of the Upper Delaware River Valley. Also, the presently available archaeological data do not seem to illuminate anything more than a band-level organization in the Middle Delaware River Valley.

The variety of social complexity observed in the Lower Delaware River Valley Late Woodland societies is interesting since all of these groups would fall within the ethnic group *Lenape*. The archaeological data from this region clearly show that there can be a great deal of variety in social organizations within a single identifiable ethnic group.

The varied social complexity observed in this area also implies a variety of cultural evolutionary pathways. In the Lower Delaware River Valley the archaeological data reveal that late Middle Woodland societies were relatively sedentary, utilized storage extensively, and were at least at a tribal level of complexity. The development of this complexity has been linked to environmental, and perhaps social, circumscription in major drainage floodplains and large estuarine marsh environments (Custer 1982a, 1982b, 1983a). These societies clearly were not supported by agriculture and are an anomaly among tribal societies. In many ways, the absence of agriculture made their complexity somewhat fragile, and once the circumscription that caused their development disappeared, as it seems to have done during early Late Woodland times, these societies apparently reverted to the more simple, nonsedentary bands that characterize the ethnographic Lenape.

The Fall Line zone of the Lower Delaware River Valley is similar to areas further south in that late Middle Woodland groups were quite complex and were not supported by agriculture (Stewart 1982a, 1982c). However, the possible addition of agriculture during Late Woodland times may have allowed this complexity to last into Late Woodland times so that larger agricultural villages came into being.

The Middle Delaware Valley is a little different in that late Middle Woodland societies in this area do not seem to have been as complex as those further south, and agriculture is a much later addition to Late Woodland subsistence systems. Also, its addition did not bring about the rise of sedentary, large villages. Perhaps this was because the addition of agriculture was very late in the Middle Delaware and the processes of population growth and community nucleation never had time to take place. In general, agriculture seems to be an important feature related to the increasing social complexity in the Middle and Lower Delaware River Valley. Because the development of agriculture in this region is most likely related to diffusion and interaction among social groups, contact with more complex groups having agriculture is a critical factor in the development, or sustaining, of initial tribal organizations.

Finally, the Lenape had matrilineal social organizations, as did the Munsee. This fact is not surprising in the areas that did have agriculture; however, their existence is more difficult to account for among the band-level societies of the southern portions of the lower Delaware River Valley. Nevertheless, if one accepts the hypothesis that matrilineal systems arise as cooperative female labor groups become important components within agricultural subsistence systems, or at least intensified gathering systems (Trigger 1978; Harris 1979, 97; Service 1962, 120–23; Martin 1974), it seems logical that matrilineal organizations developed in conjunction with the rise of sedentism and intensified plant procurement prior to Late Woodland times. They may have been retained, even after there was a decrease in cultural complexity during Late Woodland times, in the extreme southern portions of the Lower Delaware River Valley region.

Lower Delmarva Peninsula

The Lower Delmarva area, discussed by Custer and Griffith, comprises a variety of ethnic groups, including the previously discussed Lenape. As Custer and Griffith note, there is a great deal of variety in the social organizational complexity among these groups, as evidenced both in the ethnohistoric and archaeological record. Looking at the ethnohistorical data we can see that the groups living in the extreme southern portions of the Delmarva Peninsula were the most complex societies living on the peninsula. Two of these groups, the Accohanoc and the Accomac of the Eastern Shore of Virginia, were actually tributary members of the Powhatan Chiefdom, although if

their tributary status within a more complex organization is ignored, they resemble tribal organizations. As one moves north the cultural complexity tends to decrease. Along the Choptank and Nanticoke drainages and in the area of Cape Henlopen, the ethnohistoric data seem to depict tribal-level organizations. In a few cases, such as among the Choptank and possibly the Nanticoke, there are signs of some incipient supralocal organizations, which would indicate minimal chiefdoms. However, in the north the social complexity markedly declines until the Lower Delmarva Peninsula groups merge with the hunting-gathering bands of the Lenape of the northern Peninsula.

The archaeological data show a similar south-to-north gradient of indications of decreasing social complexity. In the more southern areas, there is some evidence for the use of corn agriculture, large village sites, and small ossuarylike features. However, Late Woodland societies of the more northern areas of the Lower Peninsula lack these features. Nonetheless, the Lenape of the lower Delaware Bay drainages are somewhat more complex and sedentary than the Lenape groups further to the north. Similar patterns occur among the Late Woodland societies of the western shore of the Delmarva Peninsula.

The late Middle Woodland societies of the Lower Delmarva Peninsula are the most complex Middle Woodland societies in the Middle Atlantic, and much data suggests the existence of ranked societies during Middle Woodland times (Gardner 1982; Custer 1982b). The most complex of these societies, those of the Webb complex, tend to be clustered in the central portion of the peninsula. When the Late Woodland societies of this region are compared to them there do seem to be some signs of a decrease in social complexity during early Late Woodland times. The large cemeteries and trade in exotic sumptuary goods that characterized the Webb complex ranked societies disappear during early Late Woodland times. However, the Late Woodland sites from this region are much larger and appear to be more sedentary than their Middle Woodland precursors.

The concurrent appearance of community growth and disappearance of some of the trappings of incipient ranked societies may seem somewhat anomalous because previous studies (Custer 1982b) have suggested that the disappearance of mortuary sites and regional exchange systems marks the end of ranked societies and growth in sociocultural complexity in the region due to decreased social and environmental circumscription. Nevertheless, the Late Woodland archaeological data show that even though some of the sociotechnic symbols of ranked social organizations may have disappeared, continued growth in social complexity may have occurred. Ranking may not necessarily be linked directly with other features of increasingly complex tribal societies, such as increasing local population densities, developing pan-tribal and supralocal sodalities, and sedentary lifestyles. The superstructural elements and symbols of ranked local Middle Woodland societies can thus be viewed as a somewhat ephemeral by-product of underlying changes in the infrastructure that are not necessarily directly correlated with long-term growth in sociocultural complexity.

Note too that the disappearance of trade and exchange may not even be directly related to cultural developments on the Delmarva Peninsula. For example, Stewart (1981c) notes that in the region surrounding the western Maryland and south central Pennsylvania rhyolite outcrops, which are a source for some of the Delmarva exotics, there is a distinct settlement pattern shift during Late Woodland times resulting in reduced use of the upland environments around the outcrops. Consequently, rhyolite usage is diminished and other lithic sources become important. This reduction in rhyolite usage at the quarry sources would in turn reduce the availability of surplus rhyolite for trade and exchange into the Delmarva Peninsula. Thus, the disappearance of nonlocal materials in the peninsula may not have anything to do with local sociocultural developments and, therefore, is not a good indicator of sociocultural change. It is probably safe to say that a trend of increasing social complexity leading to complex tribal organizations continued from the late Middle Woodland period into the Late Woodland period on the Lower Delmarva Peninsula.

The trend of increasing social complexity in the Lower Delmarva Peninsula can be explained on the basis of two factors. First, the large estuarine environmental settings of the lower peninsula are very productive and the intensive use of the subsistence resources from these marshes is well documented in the Late Woodland archaeological record. I suggest here that these rich coastal resources helped to support sedentary lifestyles in the face of increasing local population densities, which in turn brought about increases in sociocultural complexity. Second, and probably more important, the Lower Delmarva Peninsula was close to the complex tribal and chiefdom organizations of the western shore of the Chesapeake Bay and were undoubtedly within their interaction networks. The fact that the societies living on the southern tip of the peninsula were part of the Powhatan Confederacy supports this contention. Also, studies of ceramic designs (Griffith 1977, 1980) indicate increasing interaction with the more complex societies to the west during terminal Late

Woodland times. This interaction most likely stimulated growth in social complexity through diffusion and was probably the source of agriculture in the Lower Delmarva Peninsula.

Susquehanna Valley

The ethnohistorical data for the Susquehanna Valley, discussed by Custer in this volume, pertains almost exclusively to the Susquehannock Indians. The Susquehannocks represent a very late intrusion into the area that is probably related to the effects of European Contact. Therefore, the ethnohistorical data have little to do with the evolutionary developments of the local Late Woodland societies except in the sense that the Susquehannock brought about numerous changes in local groups' social organizations. Consequently, the ethnohistorical data is not discussed in any detail. However, the Susquehannocks represent a classic example of a lineal tribe. Their local population densities were very high and there is a good possibility that they had supralocal organizations that characterize some minimal chiefdoms. Their success in dominating other nonlineal, or cognatic, tribes shows the truth in Sahlins' 1961 arguments that segmentary lineages are highly efficient sociopolitical organizations that confer an adaptive advantage on those societies that have them.

Local Late Woodland cultures, the Shenks Ferry complex, are represented only by the archaeological record, which shows slow increases in social complexity from the Middle Woodland into the Late Woodland period leading from band-level organizations to simple, apparently cognatic, tribal organizations. Agriculture played some role in the subsistence systems, but until they came under pressure from outside groups, small isolated farmsteads and camps characterize the settlement patterns rather than sedentary villages. There is a definite causal link between increasing frequency of interactions with more complex Iroquoian lineal tribal groups to the north and the introduction of agriculture, increases in the frequency of warfare, nucleation of settlements, and the development of sedentism. There is also a clear north-to-south time slope in these developments, which further supports the notion that these social changes in the Middle and Lower Susquehanna Valley are related to interactions with Iroquoian groups. An evolutionary perspective reveals that slow growth in complexity accompanied the initial appearance of agriculture in early Late Woodland times. The rate of growth also accelerated as interactions with Iroquoian groups increased in frequency.

Thus, the degree of interaction with more complex groups seems to be the most important determinant of social complexity in the Susquehanna Valley.

Potomac Valley Coastal Plain and Piedmont

The Late Woodland cultures of the Potomac Valley were not discussed specifically in this book; however, recent reviews of the Late Woodland societies of the Potomac Piedmont and Coastal Plain are available and are briefly summarized here. The Late Woodland cultures of the Upper Potomac Valley within the Ridge and Valley and Allegheny Highland provinces are discussed in a later section.

In the recently published *Handbook of North American Indians, Volume 15, The Northeast*, the ethnohistoric data for the Lower Potomac below the Fall Line is combined with data on societies from the Delmarva Peninsula (Feest 1978b). The Upper Potomac Valley falls into a category labeled on an introductory map as "Poorly known tribes of the Ohio Valley and Interior" and is not even discussed in the text. This treatment clearly reflects the paucity of data; however, Potter (1982) provides a recent review of the ethnohistoric data that are available for the Coastal Plain areas of the Potomac drainage. He notes (1982, 35–46) that the ethnohistoric data seem to indicate that a series of "petty chiefdoms" existed in the Lower Potomac Valley at Contact; a similar assessment of social complexity is provided by Feest (1978b, 240–42). For the purposes of analysis here, Potter's petty chiefdoms are synonymous with Carniero's minimal chiefdoms described earlier. Ethnohistoric data also seem to indicate that these relatively complex organizations were supported by agriculture to some degree and stored corn and some coastal subsistence resources. It is not clear if the economies were truly redistributive, but they do seem to have been characterized by supralocal political organizations, thus qualifying them for the minimal chiefdom definition. Status differences also seem to be present, with paramount leaders of supralocal polities clearly identified.

Potter's research at a series of sites and Waselkov's (1982) intensive research at a single multicomponent midden on Virginia's Northern Neck provides the basis for the reconstruction of a possible evolutionary sequence during Late Woodland times. The sequence described below is taken from Potter's 1982 summary. During late Middle Woodland times a shift from small fusion-fission band-level organizations to more stable semisedentary village-based organizations seems to have taken place. The major archaeological evidence for this shift in organizational complexity is a settlement pattern change whereby small scattered midden and base-camp sites were replaced by larger midden sites that seemed to be occupied for a large portion of the year. By circa A.D. 700, agriculture seems to be added to the subsistence base and the degree of sedentism increased at the largest sites (Potter 1982, 346). At the same time, as Potter has written, "the mobility and territory of individual cultural groups inhabiting the Northern Neck became somewhat more restricted as the late Middle Woodland Period progressed" (1982, 347).

Between A.D. 900 and A.D. 1300 there are indications of an increase in the number of intermediate-sized sites and a dispersal of populations to neckland areas (Potter 1982, 350). Although it is not clearly demonstrated by the archaeological data, this settlement pattern may be associated with increased intensification of plant cultivation. By A.D. 1300, however, large sites seem established and the archaeological settlement pattern seems to approximate the ethnohistoric pattern characterizing minimal chiefdoms (Potter 1982, 353). Agricultural food production seems to become increasingly intensified. Note that Potter's data is derived from a single intensive sample location; however, somewhat similar settlement patterns have been noted for adjacent areas (Clark 1977, 1980; Wright 1973; L. C. Steponaitis 1980, 1983; Wanser 1982) and a similar developmental trend can be hypothesized for other Coastal Plain areas within the Potomac drainage.

Thus, in the Potomac Coastal Plain there is a trend toward increasing social complexity beginning at a band, or simple tribal, level in late Middle Woodland times that culminated in minimal chiefdoms. This growth in complexity is fueled primarily by increasingly intensified agriculture, which is systemically linked to local population growth, increased sedentism, and use of storage. Potter's comments on the decreasing sizes of group territories noted earlier may be interpreted as a description of social circumscription. If circumscription is added to the increasing population densities, a social environment comes into being that favors development of increased social complexity (Binford 1983, 208–13).

Because there are virtually no ethnohistoric data on the cultures of the Potomac Piedmont and inner Coastal Plain, one must rely strictly upon archaeological data. Clark (1977, 1980) has provided summaries of the Maryland Piedmont in the Potomac region, and although his analyses use overly simplistic equations of pottery styles and ethnic groups (see Potter 1982, 135; Potter 1980 for a discussion), Clark does show trends toward increasing village nucleation, increased reliance on agriculture, and increased incidence of warfare (as evidenced by the appearance and elaboration of village fortifications). Similar trends were noted in an earlier and more cursory overview by Schmitt (1952). Most likely these societies represent tribal organizations and are somewhat similar to the Shenks Ferry groups to the north along the Susquehanna River. It is impossible to be more specific on the processes of cultural change for this region.

Kavanagh's 1982 survey of the Monocacy Valley provides a good set of Late Woodland data for the interior Potomac Piedmont. He (1982, 69) notes three major Late Woodland complexes in the Monocacy Valley: the Montgomery complex, the Mason Island complex, and the Luray focus (see Schmitt 1952; Slattery, Tidwell, and Woodward 1966; MacCord, Schmitt, and Slattery 1957; McNett and Gardner n.d.; Manson, MacCord, and Griffin 1944 for the original complex descriptions). These complexes differ slightly in their distribution through the Monocacy Valley through time, but Kavanagh (1982, 79) notes that they all share the characteristics of large permanent or semipermanent villages, maize agriculture, defensive stockades, and a series of outlying support or procurement sites. There is an increase in the prevalence of stockade defensive structures through time as well as an intensification of floodplain utilization.

In general, the data from the Potomac Piedmont show the association of sedentism, agricultural intensification, and warfare. Although there has been no attempt to note changing interaction patterns, Kavanagh (1982, 82) reports some Iroquoian-like ceramics found in the area. Thus it might be possible that some of the complexity seen among Late Woodland groups in the Potomac Piedmont may have been stimulated by interactions with more complex societies from outside the area. However, more research will be necessary to test this hypothesis.

Virginia Coastal Plain

Turner's summary of ethnohistoric and archaeological data presented in this volume clearly shows the existence of ranked chiefdom societies in the Virginia Coastal Plain. Although the archaeological data on developing complexity is not completely clear, Gardner's 1982 study of Middle Woodland groups and Turner's data suggest an intensification of agriculture, settlement nucleation, and sedentism throughout the late Middle Woodland and Late Woodland periods. Most likely the processes of sociocultural development were similar to those described previously for the Potomac Coastal Plain. However, in the Virginia area the processes seem to have begun earlier and progressed further. This is probably due to the fact that the Virginia Coastal Plain is closer to the more complex agricultural chiefdoms of the Southeast (Hudson 1976; Swanton 1946).

Late Woodland societies of the Virginia Piedmont were clearly less complex than those of the Virginia Coastal Plain. In a recent review of the Virginia Piedmont data, Mouer (1981) notes the existence of a Monocan "confederacy," which implies a simple chiefdom organization, in the James River region. However, the term *confederacy* is misleading and Mouer's data on the Monocan societies are better characterized as lineal tribes. In this sense they are similar to Late Woodland societies of the Potomac and Susquehanna Piedmont.

Appalachian Highlands

The Appalachian Highlands are similar to many of the Piedmont areas of the Middle Atlantic in that the discussion of developing social organizations in these areas is hampered by the virtual absence of ethnohistoric data. Also, internal chronologies within the region remain undeveloped, thus making diachronic analysis difficult. Nonetheless, some relatively accurate statements about Late Woodland social systems can be made.

Most of the Late Woodland societies of the Appalachian Highlands region can be characterized as cognatic tribes. Recent summaries of local archaeological data (Wall 1981, 29–33, 134–37; R. M. Stewart 1980, 384–410; R. M. Stewart 1983, 65–69; Hughes and Weissman 1982; Cunningham, Barse, and Gardner 1979; Cunningham 1983; Carr 1983; Gardner and Boyer 1978; Boyer 1982; Custer 1979a; 1980; Foss 1983; Tolley 1983; Geier and Boyer 1982, 83–124; Geier 1979; Geier 1983; Gardner et al. 1976; Robertson and Robertson 1978; Gardner 1979; Gardner and Carbone n.d.) all note the existence of sedentary, or at least semisedentary, villages supported by agriculture in major floodplains. Stockades are often present and indicate some intergroup conflicts. These types of sites seem to date to at least the post-A.D. 1200 portion of the Late Woodland period. New data from the work of William M. Gardner and his associates in the vicinity of Front Royal, Virginia, indicate that some form of semisedentary community may have existed in the major drainage floodplains during early Late Woodland and possibly late Middle Woodland times. Furthermore, Custer (1980, 24; 1979b) has suggested that the Late Woodland and earlier Early and Middle Woodland floodplain settlements have their base in Late Archaic settlement patterns. Thus, the most that can be said about the changes through time in Late Woodland settlement patterns is that there seems to be a continued intensification of agricultural subsistence systems, culminating in settled village life in major drainage areas.

Some interregional differences in settlement patterns are significant. In a comparison of Blue Ridge and Great Valley settlement patterns, Custer (1979b; 1984) notes that in the higher elevation settings, smaller semisedentary village/base camps are found in upland areas along limited floodplains of low-order drainages. These types of sites are absent in areas with lower relief and elevation. These differences in settlement patterns are related to smaller and fewer productive floodplain settings in areas of high relief and elevation. Different social organizations may be present in each type of area; however, at present there are insufficient data to test this hypothesis. Nonetheless, in the higher elevation areas of southwest Virginia, eastern Tennessee, and western North Carolina, there appear to be incipient chiefdoms (Gardner 1979; Turner 1981) that do not occur in lower-elevation floodplains further north in the Appalachian Highlands. However, the chiefdoms noted above are also geographically closer to the chiefdoms of the upland areas of the Southeast (Dickens 1978), and it is not clear whether environmental structure or interregional interactions are responsible for the presence of these more complex social organizations.

In sum, the Late Woodland societies of the Appalachian Highlands appear to be primarily agricultural cognatic tribes living in sedentary villages. The development of these societies during the Late Woodland period seems to be related to agricultural intensification.

Peripheral Areas

In order to more fully understand the evolutionary changes occurring during Late Woodland times in the Middle Atlantic, it is useful to consider Late Woodland societies in areas adjacent to the Middle Atlantic. Three such areas are the southeastern culture area, Monongahela cultures of the upper Ohio Valley, and Iroquoian cultures of New York and western New England.

The southeastern area was briefly mentioned in the discussion of the Late Woodland cultures of the Virginia Coastal Plain, where it was noted that the chiefdoms of the southern Middle Atlantic Coastal Plain were strongly influenced by the complex Mississippian chiefdoms of the southeast. Although detailed description of the evolution of these chiefdoms is not presented here, it can be noted that the process of evolution of these chiefdoms is based on intensification of agriculture, pronounced nucleation of settlement, sedentism, and the rise of rather elaborate ceremonial centers (Muller 1983). Growth in complexity continues through most of the Woodland period and culminates in the complex Mississippian chiefdoms (Hudson 1976). Local environmental and social factors produce considerable variation in these chiefdoms (see I. F. Smith 1978). However, their developmental histories generally conform to agricultural intensification models, such as those proposed by Harris (1979, 92–100). The important point for the study of Middle Atlantic Late Woodland societies is that the only typical chiefdoms found in the Middle Atlantic, those of the Powhatan Confederacy, are the result of extensions of the growth of these more complex southeastern societies. Specifically, the growth of the Virginia Coastal Plain societies is fueled by the processes of agriculture intensification that were most likely brought into the southern Middle Atlantic through interactions within southeastern groups.

The Monongahela cultures of the Upper Ohio Valley and adjacent drainages of southwestern Pennsylvania are somewhat similar to southern Virginia Late Woodland cultures of the Middle Atlantic in that the Monongahela also are located on the periphery of Mississippian societies and apparently were greatly influenced by them. Although George (1980) has attempted to demonstrate culture links to eastern Siouan speakers, there are no ethnohistorical data on these groups (Mayer-Oakes 1955, 220–28). Archaeological data reveal that Monongahela groups participated in Mississippian interaction networks via contacts with more complex Fort Ancient groups to the west (Essenpreis 1978; Graybill 1980). Tribal-level organizations seem to have characterized most Monongahela groups, and corn agriculture, fortified sedentary villages, and intensive use of storage are common (see Herbstritt 1983, 1981; George 1983, 1980; Griffin 1983, 292–94; and Wall 1981 for area summaries). Some interaction with groups to the east and north has been suggested (George 1980, 1983; and Custer, this volume) in addition to the networks that included Fort Ancient groups. The growth in complexity to tribal levels seems to have been a process that began in Middle Woodland times (Holstein 1979), and the overall configuration of these cultures and their growth are similar to tribal societies of the Maryland and Virginia piedmont.

Iroquoian groups living to the north of the Middle Atlantic are a third peripheral group. In this volume Custer and Kraft note the importance of interactions with Iroquoian groups for growth of social complexity in the Susquehanna and Delaware drainages throughout the later half of the Late Woodland period. Ethnohistorical data

(Fenton 1978) and archaeological data (Ritchie 1969, 300–324; Ritchie and Funk 1973, 165–332, 359–69; Snow 1980, 308–18; Funk and Rippeteau 1977; Tuck 1978) clearly show the Iroquoian groups to be lineal tribes with sedentary villages, corn agriculture, and extensive use of storage. Development of these organizations has been discussed specifically by Trigger (1981), as well as in passing by others (Funk 1983, 348–62; Funk and Rippeteau 1977, 33–34, 49–50; Whallon 1968), and seems to be directly related to increasing agricultural intensification, sedentism, and nucleation of communities from late Middle Woodland through early Late Woodland times. However, although cultigens were probably originally brought into the Iroquoian area via the Mississippi and Ohio valleys during Middle Woodland times, the ensuing cultural development of Iroquoian groups was relatively independent of southern interactions until middle Late Woodland times, when Iroquoian interaction spheres shifted southward.

In sum, there were three avenues of interaction or diffusion from two major sources that drew Middle Atlantic Late Woodland groups into interaction with the more complex lineal tribes and chiefdoms that brought agriculture into the region. One source of these interactions was the southeast/Mississippian cultures, whose influences came into the Middle Atlantic from the south along the Atlantic Coastal Plain and from the west along the Ohio Valley. The second source was Iroquoian cultures, whose influences came from the north along the Susquehanna and Delaware drainages.

Late Woodland Sociocultural Evolution in the Middle Atlantic

The above summaries allow comparisons of internal social structure and evolutionary developments within the Late Woodland groups of the Middle Atlantic, as well as a study of the effects of varied external influences.

We can conclude that the degree of *agricultural* subsistence intensification and the extent of interaction with more complex societies are the two major variables that determine the extent to which Middle Atlantic Late Woodland societies grew in sociocultural complexity. These two variables are also closely related. Although some Early and Middle Woodland groups of the Middle Atlantic practiced intensive harvesting of wild seed plants (see Custer, Stiner, and Watson 1983 for an example), which could be viewed as the early stages of the processes of independent evolution of agriculture, the maize agriculture that supported many Middle Atlantic Late Woodland societies was most likely diffused into the area from agricultural societies in adjacent areas. Therefore, agricultural intensification could not have occurred until there was sufficient outside interaction to bring the components of agricultural systems into the area, even if some subsistence intensification had already taken place.

Because the variables of interaction and agricultural intensification are closely related, it is also difficult to separate their effects upon social groups. As Sahlins (1958 ix–x) notes, the cultures observed at any point in time are products of both external influences and internal evolutionary changes. For the Middle Atlantic it is possible, nonetheless, to separate the processes to a certain extent. Because there is no independent development of maize agriculture, its presence among Late Woodland societies in the Middle Atlantic is good evidence for external interactions. Ceramic design simlarities may also measure the extent of intergroup interactions (Plog 1980), and examples have been used earlier in this volume. Differential internal processes of agricultural intensification can also be discerned as separate from external interactions. For example, if the dominant community type within a region is a series of agricultural farmsteads rather than nucleated villages, it is highly likely that the process of intensification, in which population concentration and intensive use of agriculture are tightly linked (Harris 1979, 85–88; Cohen 1977; Service 1962, 111–13; Price 1982, 730–33; Price 1981), had not taken place to a large extent.

Comparing the archaeological data from the various Middle Atlantic Late Woodland societies in light of the variables of agricultural intensification and interaction makes it possible to trace out their varied evolutionary pathways, or trajectories. The approach used here is modeled after the work of Sahlins (1958), Sanders and Webster (1978), and Friedman (1982). Figure 17 shows the various evolutionary trajectories.

Beginning in Late Middle Woodland times, (circa A.D. 900) two basic types of social organizations are noted in figure 17: simple and complex bands. All of the Middle Woodland societies of the Middle Atlantic, except for Webb complex societies of the Lower Delmarva Peninsula and some groups living in the Fall Line zones of the Delaware, Susquehanna, Potomac, Rappahannock, York, and James rivers, and around Dismal Swamp, were simple bands. The exceptions were complex bands, possibly very simple tribes. At some point during the Late Woodland period these groups all experienced some kind of interaction with other agricultural groups from outside the Middle Atlantic, which made it possible for agriculture food production systems to be used. In some cases, agriculture was never added to the subsistence base. The Lenape were nonagricultural and marginal

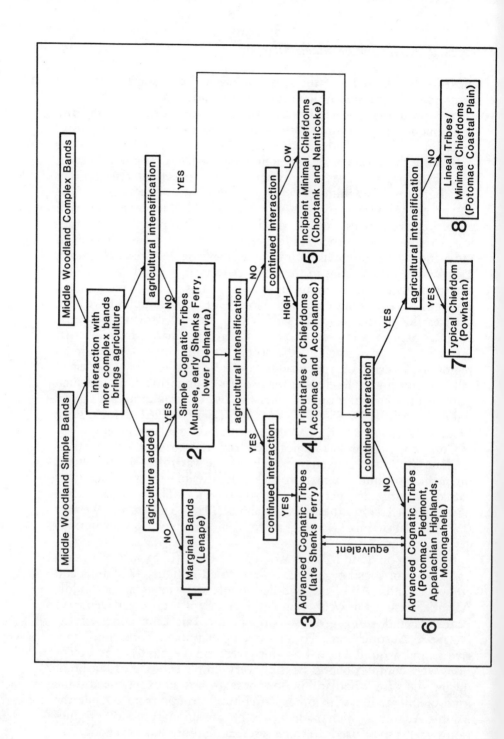

bands, which resulted from this set of events (ex. 1, fig. 17). Probably only simple bands would have followed this evolutionary path.

Other simple bands added agriculture to their subsistence base to a sufficient degree to allow the development of semisedentary farmsteads, but did not experience the development of increased intensification, which is linked with the development of villages. These groups are here termed simple cognatic tribes; the Middle Atlantic Late Woodland examples derived from a simple band organization would be the Upper Delaware groups (including the ethnohistoric Munsee—see ex. 2, fig. 17).

Archaeological evidence seems to show that almost all of the complex bands of the Middle Atlantic added agriculture to their subsistence base at some point in the Late Woodland. More important than its addition is the degree of intensification that took place. Intensification, in Price's (1981) sense of the term, refers to an increased investment of time and energy in a particular portion of the subsistence system. In the Middle Atlantic Late Woodland archaeological record, intensification is manifest in the form of increased use of storage facilities, greater complexity of storage technologies, increasingly visible archaeological evidence of cultigens, and a decrease in the range of wild food sources utilized. Related developments are increased residential stability (sedentism) and increased local population density. Although the topic is open to some debate, it will be assumed here that agricultural intensification is a by-product of sedentism and increasing local population density (see Binford 1983, 195–232).

If there is no agricultural intensification after the initial adoption of agriculture by complex bands, the resulting social form would be the simple cognatic tribe. Examples would include the early Shenks Ferry groups (Blue Rock and Stewart phase) and the majority of the Late Woodland societies of the Lower Delmarva Peninsula (ex. 2, fig. 17). It is interesting to note that simple cognatic tribes can apparently develop from either simple or complex bands. The important feature of their development is the addition of simple agriculture without a high degree of subsequent intensification. Nonetheless, in some cases in the Middle Atlantic simple cognatic tribes evidently formed a base for the development of more complex social forms. The best examples of this kind of development are the Shenks Ferry cultures of the Susquehanna Valley. Early Shenks Ferry groups are simple cognatic tribes; however, later groups (Lancaster-Funk phases) have settled villages and a greater reliance on cultigens. These developments are clearly related to agricultural intensification and continual interaction with hostile, more complex, nonlocal groups. The result is a series of fortified, at least semisedentary villages supported by agriculture—societies that are here labeled advanced cognatic tribes.

In the Lower Delmarva Peninsula region, growth in social complexity from a simple cognatic tribal base also occurred. However, it differs from the Shenks Ferry example in that there was no agricultural intensification in the Lower Delmarva Peninsula region. Instead, intensive interaction with nonlocal groups drew groups into positions of tributaries, such as in the case of the Accohannoc and the Accomac (ex. 4, fig. 17), who were tributaries to the Powhatan Chiefdom. These interactions also caused the development of shifting supralocal alliances through kinship, which characterize incipient minimal chiefdoms such as in the case of the Choptank and the Nanticoke (ex. 5, fig. 17). In these cases the proximity to the emerging chiefdom is critical. The Accohannoc and Accomac, who are closest to the Powhatan Chiefdom, became tributary units within the chiefdom while the Choptank and Nanticoke merely developed supralocal alliances. Furthermore, the Nanticoke, who are closer to the Powhatan Confederacy than the Choptank, have more clearly developed supralocal institutions and more clearly defined status positions than the Choptank. Thus, in some special cases the development of increased social complexity from a simple cognatic tribal base is simply a secondary reaction to contact with more complex societies.

In the Middle Atlantic, not all complex bands followed the trajectory leading to simple cognatic tribal level. If agricultural intensification took place, villages seem to have developed rather rapidly in the Potomac Piedmont, Appalachian Highlands, and Monongahela region, with the resulting advanced cognatic tribes (ex. 6, fig. 17) identical to those noted above for late Shenks Ferry groups. When continued interaction with complex societies is coupled with agricultural intensification, even more complicated organizations developed such as the Powhatan Chiefdom, the only typical chiefdom in the Middle Atlantic (ex. 7, fig. 17). If interaction continued without intensification, lineal tribes (or minimal chiefdoms) developed, such as the societies of the Potomac Coastal Plain (ex. 8, fig. 17).

In sum, Figure 17 illustrates the variety of social organizations that developed during Late Woodland times in the Middle Atlantic. The varying effects of agricultural intensification and interaction with more complex societies produced several evolutionary trajectories. Nevertheless, some convergences of trajectories are apparent. Whether the initial social form is a simple or a complex band level, the addition of agriculture with subsequent intensification led to simple cognatic tribes. Similarly, agricultural intensification and no interaction can lead to advanced cognatic tribes. The effects of intensification also seem to supercede those of group interaction in the formation of advanced cognatic tribes. Neither does the complexity of the initial social organization have a great effect, as shown by the fact that complex bands, simple bands, and simple cognatic tribes can develop into advanced cognatic tribes. Intensification does become important, however, to continued growth of social complexity. By itself, continued interaction of varying degrees led to any of the following social organization forms: lineal tribes, minimal chiefdoms, incipient minimal chiefdoms, or tributaries of chiefdoms. When coupled with agricultural intensification beyond that associated with the advanced cognatic tribe, interaction led to the development of typical chiefdoms.

Some Implications for Evolutionary Theory

The particular features of the evolutionary trajectories noted above have some implications for general sociocultural evolutionary theory. The most direct significance is for Fried's (1975) arguments concerning the origins of tribal-level organizations. Fried (1975, 99–102) suggests that the vast majority of tribal societies were "secondary tribes." The term *secondary* implies that tribes develop in response to extrasocietal pressures, a usage that is derived from Fried's 1960 study of the processes of state formation. These developments, which are reactions to, or effects of, interactions with other more complex societies, are contrasted with societies that grow in complexity because of internal adjustments of social systems related to factors such as population growth, agricultural intensification, or sedentism. In the Middle Atlantic, simple and advanced tribes would correspond to Fried's (1975, 99) secondary tribes. These organizations also are similar to Kroeber's (1955, 1963) notion of the "tribelet" and do not seem to be part of the main stream of cultural evolution. Once groups become cognatic tribes they seem destined to become marginal participants in more complex social systems. The fact that interaction with more complex societies is an important variable in the development of cognatic tribes in the Middle Atlantic tends to support Fried's arguments.

Lineal tribes are a different matter. Agricultural intensification is more important in the development of Middle Atlantic Late Woodland lineal tribes, although interactions with more complex societies were still important. Intensification of agriculture, increased sedentism, and population growth were all linked so that more efficient lineage organizations had adaptive values (Sahlins 1961; Kristiansen

1982). These lineages, in their various forms, in turn provided the social relationships that developed into the supra-local socio-political organizations characterizing minimal and typical chiefdoms. Thus, lineal tribes are more likely to be precursors of chiefdoms than are cognatic tribes.

In some ways, the evolutionary trajectories described for various Middle Atlantic Late Woodland societies and the general statements about the evolutionary potential of cognatic and lineal tribes made here parallel conclusions made by Sahlins (1958, 201–3) concerning evolving social systems in Polynesia. Sahlins (1958, 139–96) distinguishes between two types of organizations: ramified and descent-line. The ramified systems are analagous to the lineal tribes described here in their basic organizations although the ramified Polynesian societies are clearly chiefdoms. Descent-line systems, on the other hand, are analagous to cognatic tribes in their basic organizations. In the Sahlin's Polynesian examples, ramified systems are generally more complex than the descent-line systems. Similarly, the Middle Atlantic Late Woodland lineal tribes were more likely to develop into more complex organizations than the cognatic tribes.

In considering the differences between the ramified and descent-line systems and the origins of each type of system, Sahlins (1958, 201–3) suggests a general evolutionary pattern that also has some implications for the development of cognatic and lineal tribes. He notes:

> Apparently, the crucial difference between ramified and descent-line systems is whether or not kinship relationship is maintained between households of patrilocal extended families which have segmented because of increase in size. In both organizations, large groups of this nature divide, which gives rise to two distinct alignments of patrilineal kin. *In ramified organizations, the split usually takes place between the families of two brothers. The relationship is maintained, and the two households are ranked according to seniority of the brothers. In descent-line systems, the connection between the segmented families may be remembered for a time, but gradually the tie weakens and two discrete groups are formed.* Practically all other differences in the two systems are related to this difference in the segmentation process. (1958, 201–2, Sahlin's emphasis)

The most important feature of the contacts maintained among unilineal kin groups is economic cooperation, and Sahlins further suggests that the maintenance of kinship ties, which characterizes ramified systems, is "an adaptation to the economic necessities of cooperation" (1958, 202). If multiple households that have resulted from fissioning of a common source each begin to have a degree of specialization of production, either by choice or by necessity, the cooperative production, and therefore ramified systems, are more likely to develop. Furthermore, Sahlins (1958, 203) suggests that ramified systems are adapted to the exploitation of different, widely spread resource areas. If fissioning causes groups to move into areas with similar resources, the adaptive value of cooperative production is reduced and descent-line systems are more likely to develop. The general evolutionary model might be stated as follows:

> In the face of fissioning of households caused by population growth, the probability of the development of ramified systems is directly proportional to the degree of resource production specialization and inversely proportional to local habitat diversity. Conversely, the probability of the development of descent-line systems is inversely proportional to the degree of resource production specialization and directly proportional to local habitat diversity. (Sahlins 1958; 202)

In order to apply this model to the Middle Atlantic Late Woodland examples, cognatic tribes can be substituted for descent-line systems and lineal tribes can be substituted for ramified systems. The Middle Atlantic examples tend to follow the same pattern of evolutionary potential. The examples of cognatic tribal organizations noted in figure 17 tend to be series of agricultural villages that are located in floodplain settings. These settings, with their rich agricultural soils, tend to be relatively homogeneously distributed along the major drainages in the areas where cognatic tribes are found. On the other hand, the lineal tribes seem to develop where productive settings are more widely spaced, such as in Coastal Plain areas where poorly drained and very dry soils of low agricultural productivity separate areas with highly productive soils. In addition, the use of coastal and estuarine resources, which are not uniformly distributed, creates a setting where local subsistence specialization is more likely to occur. In Coastal Plain areas such as the Virginia Tidewater, economic specialization, at least to a limited degree, may have made intergroup cooperation more likely and created a setting where lineal tribal organizations, and perhaps ultimately ramified social systems, conferred an adaptive advantage. However, a complete test of the applicability of Sahlins' hypothesis must await the development of less impressionistic methods for measuring the structure of environmental distributions and their correlation with local settlement patterns.

The importance of internal agricultural intensification for formation of lineal tribes in the Middle Atlantic, and eventually for minimal and typical chiefdoms, suggests that theories on the origins of chiefdoms that stress other factors may need some revision. For example, Carniero (1981, 54–65) notes that warfare is the major cause of the development of chiefdoms. In the Middle Atlantic, interactions of lineal tribes with other lineal tribes and more complex chiefdoms most likely did include warfare, as evidenced by the presence of stockades and village fortifications. However, agricultural intensification created the social settings within which this warfare took place. Therefore, if one follows the evolutionary trajectories that produce chiefdoms back to the level of lineal tribes, factors other than warfare must be considered.

As a final note, it is useful to compare Late Woodland cultural evolution to Late Archaic–Middle Woodland cultural developments. During Late Archaic–Middle Woodland times, coastal-adapted societies, especially those living on the Delmarva Peninsula, were the most complex in the Middle Atlantic (Custer and Stewart 1983). The effects of circumscription on societies in rich coastal environments are dominant factors causing the development of ranked societies. However, during Late Woodland times, the coastal-adapted societies do not add intensive agriculture to their subsistence systems and grow in complexity much more slowly than do those other groups who had more simple organizations during Middle Woodland times but who do adopt intensive agriculture during Late Woodland times. Thus, the specific nature of the coastal adaptions, which produced a rapid yet ephemeral growth of ranked societies in the Middle Atlantic, precluded the possibility for further sustained growth supported by agriculture. Similar patterns are noted in other areas of the world (Sahlins and Service 1960, chap. 5; Sanders and Webster 1978).

In conclusion, the Late Woodland societies of the Middle Atlantic show a great deal of diversity in their cultures and their evolutionary histories. However, the common themes of evolutionary development and convergences of evolutionary trajectories can contribute insights into how and why the processes of cultural evolution work.

References

Abbott, C. C.
   1872   The stone age in New Jersey. *American Naturalist* 6:144–60, 199–229.

Alexander, J.
   1969   The indirect evidence for domestication. In *The domestication and exploitation of plants and animals*, ed. P. J. Ucko and G. W. Dimbleby, 123–30. New York: Aldine.

Ameringer, C.
   1975   Susquehannock plant utilization. In *Proceedings of the 1975 Middle Atlantic Archaeological Conference*, ed. W. F. Kinsey, 58–63. Lancaster, Penn.: North Museum, Franklin and Marshall College.

Anderson, C.
   1973   Organic and stone materials from the Heisey collection of Strickler site artifacts. North Museum, Franklin and Marshall College, Lancaster, Pennsylvania. Typescript.

Anonymous
   1951a   Aboriginal evidence from the grounds of the Lewes School. *Archeolog* 3(1):3–4.
   1951b   The Derrickson site "worked" conchs. *Archeolog* 4(1):9–16.

Arber, E., ed.
   1910   *The travels and works of Captain John Smith.* Edinburgh: John Grant.

Artusy, R. E.
   1976   An overview of the proposed ceramic sequence for southern Delaware. *Maryland Archaeology* 12(2):1–15.

Artusy, R. E., and D. R. Griffith
   1975   A brief report on the semi-subterranean dwellings of Delaware. *Archeolog* 27(1):1–9.

Baker, J. A.
   1980   The economics of weed seed subsistence in the Ridge and Valley Province of central Pennsylvania. In *The Fisher Farm site: A Late Woodland hamlet in context*, 205–22. Pennsylvania State University Department of Anthropology Occasional Papers in Anthropology, ed. J. W. Hatch, no. 12. University Park.

Barbour, P. L., ed.
   1969   *The Jamestown Voyages under the first charter 1606–1609.* London: Cambridge Univ. Press.

Barka, N. F., and B. C. McCary
   1969   The Chicahominy River survey of eastern Virginia. *Proceedings of the Eastern States Archaeological Federation* 26:20.

Bastian, T.
   1974   The 19th Maryland archaeological field session, the Wessel site (18CA21), Tuckahoe State Park, Caroline County, Maryland. Division of Archaeology, Maryland Geological Survey, Baltimore. Typescript.

Becker, M. J.
   1976   The Okehocking: A remnant band of Delaware Indians in Chester County, Pennsylvania, during the colonial period. *Pennsylvania Archaeologist* 46(3):25–63.
   1980a   Lenape archaeology: Archaeological and ethnohistoric considerations in light of recent excavations. *Pennsylvania Archaeologist* 50(4):19–30.
   1980b   Wampum: The development of an early American currency. *Bulletin of the Archaeological Society of New Jersey* 36:1–11.
   1982   The Okehocking band of Lenape: Cultural continuities and accommodations to colonial expansion in the early 18th century. Paper presented by the annual meeting of the American Society for Ethnohistory, Nashville, Tennessee.
   n.d.a.   The boundary between the Lenape and the Munsee: Indications that the Forks of the Delaware was in a buffer zone during the early historic period. In *Man in the Northeast.* In press.
   n.d.b.   The Lenape of southern New Jersey: Cultural identity and continuity during the colonial period. *New Jersey History.* In press.
   n.d.c.   The Lenape southern boundary: Cultural interaction and change in the early contact period, 1550–1610. Paper presented at the Laurier II Conference, Huron College, London, Ontario.
   n.d.d.   Teedyuscung's land rights near Toms River, New Jersey: The cultural boundaries of the Jersey Lenape and their movement into Pennsylvania as deduced from a land sale in 1734. Typescript.
   n.d.e.   Hannah Freeman, 1730–1801: The Life of a Lenape woman in early Pennsylvania as an example of one mode of accommodation to colonial expansion. Typescript.
   n.d.f.   Penn's protection of Lenape land rights: Efforts to protect native claims to lands *de facto* reserved for their use within the tracts sold to the proprietary government of Pennsylvania. Typescript.
   n.d.g.   The Okehocking: Cultural continuities within a Lenape band in southeastern Pennsylvania in the early 18th century. Typescript.
   n.d.h.   The Montgomery site, 36CH60: A late contact Lenape (Delaware) site in Wallace Township, Chester County, Pennsylvania. Anthropology Section, West Chester University, West Chester, Pennsylvania. Typescript.

Bernabo, C.
 1981 Quantitative estimates of temperature change over the past 2700 years in Michigan based on pollen data. *Quaternary Research* 15:143–59.

Beverly, R.
 1947 *The history and present state of Virginia.* Chapel Hill: Univ. of North Carolina Press.

Binford, L. R.
 1964 *Archaeological and ethnohistorical investigation of cultural diversity and progressive development among aboriginal cultures of coastal Virginia and North Carolina.* Ann Arbor, Mich.: University Microfilms.
 1983 *In pursuit of the past: Decoding the archaeological record.* New York: Thames and Hidson.

Bishop, S. C.
 1935 Fisheries investigations in the Delaware and Susquehanna rivers. In *Biological survey of the Delaware and Susquehanna watersheds,* 122–39. Supplement to the 25th Annual Report of the New York State Conservation Department, Section VI. Albany.

Blaker, M. C.
 1963 Aboriginal ceramics from the Townsend site. *Archeolog* 15(1):14–39.

Boyer, W. P.
 1982 *Archaeological survey and site location prediction: towards an understanding of prehistoric settlement in western Virginia.* Ann Arbor, Mich.: University Microfilms.

Braun, D. P., and S. Plog
 1982 Evolution of "tribal" social networks: Theory and prehistoric North American evidence. *American Antiquity* 47:504–25.

Bressler, J. P.
 1975 Excavation of a rockshelter on Grays Run, Lycoming County, Pennsylvania (36LY134). Lycoming County Historical Society, Williamsport, Pennsylvania. Typescript.
 1980 Excavation of the Bull Run site (36LY119). *Pennsylvania Archaeologist* 50(4):31–63.

Brinton, D. G.
 1885 *The Lenape and their legends, with the complete text and symbols of the Walam Olum.* Philadelphia: Privately printed.

Brown, J. A.
 1971 The dimensions of status burials at Spiro. In *Approaches to the social dimensions of mortuary practices,* 92–112. Society for American Archaeology Memoir no. 25. Washington, D.C.

Browne, W. H., ed.
1887 *Proceedings of the Council of Maryland 1667–1687/8.* Baltimore: Archives of Maryland.

Bryant, C. L., R. C. Rosser, H. Hutchinson and H. Hutchinson
1951 The Willin Farm site. *Archeolog* 3(3):1–3.

Buchanan, W. T.
1966 The Camp Weyanoke site, Charles City, Virginia. *Quarterly Bulletin of the Archaeological Society of Virginia* 21:42–48.
1969 The Deep Bottom site, Henrico County, Virginia. *Quarterly Bulletin of the Archaeological Society of Virginia* 23:103–14.
1976 The Hallowes site, Westmoreland County, Virgina. *Quarterly Bulletin of the Archaeological Society of Virginia* 30:195–99.

Budd, T.
1685 *Good order established in Pennsylvania and New Jersey, being an account of the country; with its produce and commodities then made in the year 1685.* New York.

Bushnell, D. I.
1920 *Native cemeteries and forms of burial east of the Mississippi.* Bulletin of the Bureau of American Ethnology 71. Washington, D.C.
1937 *Indian sites below the falls of the Rappahannock, Virginia.* Smithsonian Miscellaneous Collections 96(4). Washington, D.C.
1940 *Virginia before Jamestown.* Smithsonian Miscellaneous Collections 100. Washington, D.C.

Cadzow, D. A.
1936 *Archaeological studies of the Susquehannock Indians of Pennsylvania.* Safe Harbor Report no. 2. Harrisburg: Pennsylvania Historical and Museum Commission.

Callaway, W., H. Hutchinson, and D. Marine
1960 The Moore site. *Archeolog* 12(1):1–8.

Cameron, L. D.
1976 Prehistoric hunters and gatherers of the upper Chesapeake Bay region: A study on the use of a predictive model for the analysis of settlement-subsistence systems. Undergraduate honors thesis, Department of Anthropology, University of Michigan, Ann Arbor.

Carbone, V. A.
1976 *Environment and prehistory in the Shenandoah Valley.* Ann Arbor, Mich.: University Microfilms.
1982 Environment and society in Archaic and Woodland times. In *Practicing environmental archaeology: Methods and interpretations,* ed. R. Moeller, 39–52. Occasional Papers of the American Indian Archaeological Institute no. 3. Washington, Connecticut.

Carniero, R. L.
1967   On the relationship between size of population and complexity of social organization. *Southwestern Journal of Anthropology* 23:234–43.
1981   The chiefdom: Precursor of the state. In *The transition to statehood in the New World*, ed. G. D. Jones and R. R. Kautz, 37–75. New York: Cambridge Univ. Press.

Carpenter, E. S.
1949   The Brock site. *Pennsylvania Archaeologist* 19(3–4):69–77.

Carr, K. W.
1983   A predictive model for prehistoric site distribution in Berkeley County, West Virginia. In *Upland archaeology in the East: A symposium*, ed. C. R. Geier, M. B. Barber, and G. A. Tolley, 141–70. Washington, D.C.: U.S. Forest Service.

Casselberry, S.
1971   *The Schultz-Funck site (36LA7): Its role in the culture history of the Susquehannock and Shenks Ferry Indians.* Ann Arbor, Mich.: University Microfilms.

Catlin, M., J. F. Custer, and R. M. Stewart
1982   Late archaic cultural change in Virginia. *Quarterly Bulletin of the Archaeological Society of Virginia* 37:123–40.

Cavallo, J.
1982   Fish, fires, and foresight: Middle Woodland economic adaptations in the Abbott Farm National Landmark, Trenton, New Jersey. Paper presented at the annual meeting of the Eastern States Archaeological Federation, Norfolk, Virginia.

Ceci, L.
1977   *The effect of European Contact and trade on the settlement pattern of Indians in coastal New York, 1524–1665: the archaeological and documentary evidence.* Ann Arbor, Mich.: University Microfilms.

Chittenden, M. E.
1972   Response of young shad *Alosa sapidissma* to low temperatures. *Transactions of the American Fisheries Society* 101:680–85.

Clark, W. E.
1976   An archaeological reconnaissance of the Captains Cove development. Virginia Research Center for Archaeology, Yorktown. Typescript.
1977   The Potomac Creek complex: A Late Woodland development. Paper presented at the Middle Atlantic Archaeological Conference, Trenton, New Jersey.
1980   The origins of the Piscataway and related Indian cultures. *Maryland Historical Magazine* 75(1):8–22.

Cohen, M.
1977   The food crisis in prehistory: overpopulation and the origins of agriculture. New Haven, Conn.: Yale Univ. Press.

Corkran, D. E., and P. S. Flegel
1953   Notes on the Marshyhope Creek sites (Maryland). *Archeolog* 5(1):46.

Crabtree, P. J., and A. Langendorfer
1981   Paleoethnobotany of the Delaware Park site. *MASCA Newsletter* 1(7):195–201.

Cross, D.
1941   *Archaeology of New Jersey, Volume I*. Archaeological Society of New Jersey and the New Jersey State Museum, Trenton.
1956   *Archaeology of New Jersey, Volume II: The Abbott Farm*. Archaeological Society of New Jersey and the New Jersey State Museum, Trenton.
n.d.   Lab guide to arrowhead, spearhead, drill, and scraper types. Louis Berger and Associates, Inc., East Orange, New Jersey. Typescript.

Cunningham, K. W.
1983   Prehistoric settlement-subsistence patterns in the Ridge and Valley section of the Potomac Highlands of eastern West Virginia: Hampshire, Hardy, Grant, Mineral, and Pendleton counties. In *Upland archaeology in the East: A symposium*, ed. C. R. Geier, M. B. Barber, and G. A. Tolley, 171–224. Washington, D.C.: U.S. Forest Service.

Cunningham, K. W., W. P. Barse, and W. M. Gardner
1979   A preliminary archaeological assessment of Appalachian Corridor H, West Virginia. West Virginia Geological and Economic Survey, Morgantown. Typescript.

Custer, J. F.
1978   Broadspears and netsinkers: Late Archaic adaptations at four Middle Atlantic archaeological sites. Paper presented at the Middle Atlantic Archaeological Conference, Rehoboth Beach, Delaware.
1979a   *An evaluation of sampling techniques for cultural resources reconnaissance in the Middle Atlantic area of the United States*. Ann Arbor, Mich.: University Microfilms.
1979b   Settlement-subsistence systems in the Blue Ridge and Great Valley sections of Virginia: A comparison. Paper presented at the Middle Atlantic Archaeological Conference, Rehoboth Beach, Delaware.
1980   Settlement-subsistence systems in Augusta County, Virginia. *Quarterly Bulletin of the Archaeological Society of Virginia* 35:1–27.

1981 *Report on archaeological research in Delaware by the University of Delaware Department of Anthropology (FY 1981)*. Dover: Delaware Division of Historical and Cultural Affairs.
1982a The prehistoric archaeology of the Churchmans Marsh vicinity: An introductory analysis. *Bulletin of the Archaeological Society of Delaware* 13:1–41.
1982b A reconsideration of the Middle Woodland cultures of the Upper Delmarva Peninsula. In *Practicing environmental archaeology: Methods and interpretations*, ed. R. Moeller, 29–38. Occasional Papers of the American Indian Archaeological Institute no. 3.
1983a *Delaware prehistoric archaeology: An ecological approach*. Newark: University of Delaware Press.
1983b Late Archaic and Delmarva Adena settlement patterns of central Delaware: Implications for the origins of ranked societies. Paper presented at the annual meeting of the Society for American Archaeology, Pittsburgh, Pennsylvania.
1983c *A management plan for the upper Delmarva region of Maryland, Cecil, Kent, Talbot, Queen Anne, Caroline, and upper Dorchester counties, Maryland*. Maryland Historical Trust Manuscript Series no. 31. Annapolis.
1984 A controlled comparison of Late Woodland settlement patterns in the Appalachian highlands. In *Upland Archaeology in the East, Symposium 2*, ed. M. B. Barber, 75–101. Washington, D.C.: U.S. Forest Service.
n.d. Test excavations at the Webb site (36CH51), Chester County, Pennsylvania. *Pennsylvania Archaeologist*.

Custer, J. F., and D. Bachman
1982 *Phase II archaeological investigations at 7NC-D-75, 7NC-E-43, and 7NC-E-45, New Castle County, Delaware*. Dover: Delaware Department of Transportation.

Custer, J. F., W. Catts, and D. Bachman
1982 *Phase II archaeological investigations at two prehistoric sites, 7NC-D-70 and 7NC-D-72, New Castle County, Delaware*. Dover: Delaware Department of Transportation.

Custer, J. F., and K. R. Doms
1983 *A reanalysis of the Wilke-Thompson collections, Kent County, Maryland*. Maryland Historical Trust Manuscript Series no. 30. Annapolis.
n.d. Analysis of collections from the Oxford site (18TA3), Talbot County, Maryland. *Maryland Archaeology*. In press.

Custer, J. F., and A. A. Futer
1978 Status and role in Susquehannock mortuary ceremonialism: Data from the Strickler site (36LA3). Paper presented at the annual meeting of the Society for Pennsylvania Archaeology, Champion, Pennsylvania.

Custer, J. F., and G. J. Galasso
1983　An archaeological survey of the St. Jones and Murderkill drainages, Kent County, Delaware. *Bulletin of the Archaeological Society of Delaware* 14:1–18.

Custer, J. F., J. M. McNamara, and H. H. Ward
n.d.　Woodland ceramic sequences of the Upper Delmarva peninsula and southeastern Pennsylvania. *Maryland Archaeology*. In press.

Custer, J. F., J. Sprinkle, A. Flora, and M. Stiner
1981　The Green Valley site complex: Lithic reduction base camp sites on the Delaware Fall Line. *Bulletin of the Archaeological Society of Delaware* 12:1–31.

Custer, J. F., and R. M. Stewart
1983　Maritime adaptations in the Middle Atlantic region of Eastern North America. Paper presented at the annual meeting of the Society for American Archaeology, Pittsburgh, Pennsylvania.

Custer, J. F., M. C. Stiner, and S. C. Watson
1983　Excavations at the Wilgus site (7S-K-21), Sussex County, Delaware. *Bulletin of the Archaeological Society of Delaware* 15:1–44.

Custer, J. F., and E. B. Wallace
1982　Patterns of resource distribution and archaeological settlement patterns in the Piedmont Uplands of the Middle Atlantic region. *North American Archaeologist* 3:139–72.

Cutler, H. C., and L. C. Blake
1967　Notes on plants from the Sheep Rock Shelter. In *Archaeological investigations of the Sheep Rock Shelter, Huntingdon County, Pennsylvania, Vol. I*, ed. J. W. Michels and I. F. Smith, 125–68. University Park: Pennsylvania State Univ. Department of Anthropology.

Dankers, J., and P. Sluyter
1867　*Journal of a voyage to New York and a tour in several of the American colonies in 1679–1680*. Ed. and trans. H. C. Murphy. Brooklyn, N.Y.: Long Island Historical Society.

Davidson, D. S.
1935a　Notes on Slaughter Creek. *Bulletin of the Archaeological Society of Delaware* 2(2):1–5.
1935b　Burial customs on the Delmarva Peninsula and the question of their chronology. *American Antiquity* 1(1):84–97.
1936　Notes on faunal remains from Slaughter Creek. *Bulletin of the Archaeological Society of Delaware* 2(4):29–34.

Davidson, T. E.
1982a　*A cultural resource management plan for the Lower Delmarva region of Maryland*. Maryland Historical Trust Manuscript Series no. 2. Annapolis.

1982b   Historically attested Indian villages of the lower Delmarva. *Maryland Archaeology* 18(1):1–8.

Delaware Division of Historical and Cultural Affairs
1976   National Register nomination for the Cape Henlopen archaeological district. Delaware Bureau of Archaeology and Historic Preservation, Dover. Typescript.
1978   National Register nomination for the St. Jones Neck archaeological district. Delaware Bureau of Archaeology and Historic Preservation, Dover. Typescript.
1980   National Register nomination for the Millman site complex. Delaware Bureau of Archaeology and Historic Preservation, Dover. Typescript.

De Valinger, L.
1940   Indian land sales in Delaware. *Bulletin of the Archaeological Society of Delaware* 3(3):29–31; 3(4):25–33.

De Vries, D. P.
1909   From the "Korte historiael ende journaels aenteyckeninge," by D. P. DeVries, 1633–1643 (1655). In *Narratives of New Netherlands, 1609–1664*, ed. J. Franklin Jameson, 186–234. New York: Charles Scribners.

Dickens, R. S.
1978   Mississippian settlement patterns in the Appalachian summit area: The Pisgah and Quala Phase. In *Mississippian settlement patterns*, ed. B. Smith, 115–40. New York: Academic Press.

Didier, M.
1975   The argillite problem revisited: An archaeological and geological approach to a classical archaeological problem. *Archaeology of Eastern North America* 3:90–100.

Diehl, R. A.
1973   *Political evolution and the formative period of Mesoamerica*. Pennsylvania State Univ., Department of Anthropology Occasional Papers in Anthropology no. 8. University Park.

Doms, K. R., and J. F. Custer
1983   *Analysis of oyster shell remains and other ecofacts from a Late Woodland pit feature at the Island Field site (7K-F-17), Kent County, Delaware*. University of Delaware Center for Archaeological Research Report no. 2. Newark.

Eglan, F.
1954   Social anthropology and the method of controlled comparison. *American Anthropologist* 56:743–63.

Ember, M., and C. R. Ember
1971   The conditions favoring matrilocal versus patrilocal residence. *American Anthropologist* 73:571–94.

Essenpreis, P. S.
1978   Fort Ancient settlement: Differential response at a Mississippian-–Late Woodland interface. In *Mississippian settlement patterns,* ed. Smith, 141–68. New York: Academic Press.

Eveleigh, T., J. F. Custer, and V. Klemas
1983   LANDSAT-generated predictive models for prehistoric archaeological site locations on Delaware's Coastal Plain. *Bulletin of the Archaeological Society of Delaware* 14:19–37.

Feest, C.
1973   Southeastern Algonquian burial customs: Ethnohistoric evidence. In *Proceedings of the 1973 Middle Atlantic Archaeological Conference,* ed. R. Thomas, 1–9. Penns Grove, N.J.: Middle Atlantic Archaeological Research, Inc.
1978a   Virginia Algonquians. In *Handbook of North American Indians. Vol. 15, The Northeast,* ed. B. Trigger, 253–70. Washington, D.C.: Smithsonian Institution.
1978b   Nanticoke and neighboring tribes. In *Handbook of North American Indians. Vol. 15, The Northeast,* ed. B. Trigger, 240–52. Washington, D.C.: Smithsonian Institution.

Fehr, E., F. D. Staats, E. Erb, and L. Farina
1971   The Sandt's Eddy site. *Pennsylvania Archaeologist* 41(1–2):39–52.

Fenton, W. N.
1978   Northern Iroquoian culture patterns, In *Handbook of North American Indians. Vol. 15, The Northeast,* ed. B. Trigger, 296–321. Washington, D.C.: Smithsonian Institution.

Flannery, K. V.
1972   The cultural evolution of civilizations. *Annual Review of Ecology and Systematics* 3:399–426.
1976   *The early Mesoamerican village.* New York: Academic Press.

Flannery, R.
1939   *An analysis of coastal Algonkian culture.* Catholic University of America Anthropological Series, no. 7. Washington, D.C.

Flegel, P. S.
1959   Additional data on the Mispillion site. *Archeolog* 11(2):1–16.
1975a   Final report, Lankford site (18DO43), pit 2BF1. Division of Archaeology, Maryland Geological Survey, Baltimore. Typescript.
1975b   The Lankford site, report on pit 3 excavation. Division of Archaeology, Maryland Geological Survey, Baltimore, Typescript.
1976   The Lankford site, report on pit 1 excavation. Division of Archaeology, Maryland Geological Survey, Baltimore. Typescript.
1978   The Marshyhope Creek: Its Indian places, pottery, points, and pipes. *Archeolog* 30(1):13–59.

Forks of the Delaware Chapter
  1980   The Overpeck site. *Pennsylvania Archaeologist* 50(3):1–46.
Foss, R. W.
  1983   Blue Ridge prehistory: A perspective from the Shenandoah National Park. In *Upland archaeology in the East: A symposium*, ed. C. R. Geier, M. B. Barber, G. A. Tolley, 91–103. Washington, D.C.: U.S. Forest Service.
Fried, M. H.
  1960   On the evolution of social stratification and the state. In *Culture and history: Essays in honor of Paul Radin.* ed. S. Diamond, 713–31. New York: Columbia Univ. Press.
  1967   *The evolution of political society.* New York: Random House.
  1975   *The notion of tribe.* Menlo Park, Calif.: Cummings.
Friedman, J.
  1982   Catastrophe and continuity in social evolution. In *Theory and explanation in archaeology,* ed. C. Renfrew, M. J. Rowlands, and B. A. Segraves, 175–96. New York: Academic Press.
Fuller, J. W.
  1976   The development of sedentary village communities in northern West Virginia: The test of a model. Paper presented at the annual meeting of the Society for American Archaeology, St. Louis, Missouri.
Funk, R. E.
  1983   The northeastern United States. In *Ancient North Americans.* ed. J. D. Jennings, 303–72. San Francisco: Freeman.
Funk, R. E., and B. E. Rippeteau
  1977   *Adaptation, change, and continuity in Upper Susquehanna prehistory.* Man in the Northeast, Occasional Papers in Northeastern Anthropology no. 3. Rindge, New Hampshire.
Futer, A. A.
  1959   The Strickler site. In *Susquehannock miscellany,* ed. J. Witthoft and W. F. Kinsey, 136–47. Harrisburg: Pennsylvania Historical and Museum Commission.
Galasso, G. J.
  1981   An analysis of artifacts from Delaware in the Smithsonian Institution, Museum of Natural History. Department of Anthropology, Univ. of Delaware, Newark. Typescript.
Gardner, W. M.
  1977   *Excavations at 18PR141, 18PR142, and 18PR143, Prince Georges County, Maryland.* Washington Suburban Sanitary Commission, Washington, D.C.
  1978   Comparison of Ridge and Valley, Blue Ridge, Piedmont, and Coastal Plain Archaic site distributions: An idealized transect (preliminary version). Paper presented at the annual Middle Atlantic Archaeological Conference, Rehoboth Beach, Delaware.

1979 A Phase I cultural reconnaissance of the proposed Appalachian Power Company hydroelectric projects in Poor Valley and Hidden Valley, Washington County, Virginia. Virginia Research Center for Archaeology, Yorktown. Typescript.

1981 An intensive archaeological reconnaissance of a proposed coal transshipment facility site, Portsmouth, Virginia. Virginia Research Center for Archaeology, Yorktown. Typescript.

1982 Early and Middle Woodland in the Middle Atlantic: An overview. In *Practicing environmental archaeology: Methods and interpretations*, 53–87. Occasional Papers of the American Indian Archaeological Institute no. 3. Washington, D.C.

Gardner, W. M., and W. P. Boyer
1978 *A cultural resources reconnaissance of portions of the northern segment of Massanutten Mountain in the George Washington National Forest, Page, Warren, and Shenandoah Counties, Virginia*. Washington, D.C.: U.S. Forest Service.

Gardner, W. M., and V. A. Carbone
n.d. *The prehistoric archaeology of the Middle Atlantic.* New York: Academic Press. In press.

Gardner, W. M., E. J. Ham, D. A. Hall, J. M. McNamara, D. C. Curry, S. G. Russell, and K. W. Cunningham
1976 An archaeological survey of proposed dam sites along the upper New River. Virginia Research Center for Archaeology, Yorktown. Typescript.

Gardner, W. M., and L. Rappleye
1980 Phase II archaeological reconnaissance in the Alternate C, Variation 2, Maryland Route 404 relocated, Denton By-Pass, Caroline County, Maryland. Archaeology Laboratory, Catholic Univ. of America, Washington, D.C. Typescript.

Gardner, W. M., R. Wall, G. Tolley, and J. F. Custer
1978 A partial cultural reconnaissance of Smith Island, Maryland. Archaeology Laboratory, Catholic University of America, Washington, D.C. Typescript.

Geier, C. R.
1979 *Cattle, sheep, and white-tail deer: Man in mountainous Virginia.* Vol. 3: *The cultural sequence.* James Madison University Occasional Papers in Anthropology no. 6. Harrisonburg, Virginia.

1983 Preliminary observations on aboriginal settlement on headwater streams of the James River: The Ridge and Valley province of Virginia. In *Upland archaeology in the East: A symposium*, ed. C. R. Geier, M. B. Barber, and G. A. Tolley, 225–49. Washington, D.C.: U.S. Forest Service.

Geier, C. R., and W. P. Boyer
1982 *The Gaithright Dam—Lake Moomaw cultural resource investigations: A synthesis of the prehistoric data.* James Madison

University Occasional Papers in Anthropology no. 15. Harrisonburg, Virginia.

George, R. L.
1980   Notes on the possible cultural affiliation of Monongahela. *Pennsylvania Archaeologist* 50(1–2):45–50.
1983   The Gnagey site and the Monongahela occupation of the Somerset Plateau. *Pennsylvania Archaeologist* 53(4):1–97.

Gibson, J. L.
1974   Poverty Point, the first American chiefdom. *Archaeology* 27(2):96–105.

Gilsen, L.
1979   The environmental ecology of Calvert County, Maryland, parts I and II. *Maryland Archaeology* 15(1–2):1–30.
1980   The environmental ecology of Calvert County, Maryland, part III. *Maryland Archaeology* 16(2):14–31.

Goddard, I.
1974   The Delaware language, past and present. In *A Delaware Indian symposium*, ed. H. C. Kraft, 103–10. Anthropological Series of the Pennsylvania Historical and Museum Commission no. 4. Harrisburg.
1978a Delaware. In *Handbook of North American Indians. Vol. 15, The Northeast*, ed. B. Trigger, 213–39. Washington, D.C.: Smithsonian Institution.
1978b Eastern Algonkian languages. In *Handbook of North American Indians. Vol. 15, The Northeast*, ed. B. Trigger, 70–77. Washington, D.C.: Smithsonian Institution.

Godfrey, M.
1980   *The Piedmont.* San Francisco: Sierra Club Books.

Graybill, J. R.
1973   Shenks Ferry settlement patterns in southern Lancaster County. *The Kithuwan, Journal of the Anthropology Club, Franklin and Marshall College* 5:7–24.
1980   Marietta Works, Ohio, and the eastern periphery of Fort Ancient. *Pennsylvania Archaeologist* 50(1–2):51–60.
1982   Review of *The Fisher Farm site: A Late Woodland hamlet in context*, by J. W. Hatch. *Pennsylvania Archaeologist* 52(3–4):70–71.

Green, R.
1968   A preliminary analysis of 1967 Sheep Rock features and activity areas. In *Archaeological investigations of Sheep Rock Shelter. Huntingdon County, Pennsylvania, Vol. III*, ed. J. W. Michels and J. S. Dutt, 181–98. University Park: Pennsylvania State Univ. Department of Anthropology.

Gregory, L. B.
- 1980 The Hatch site: Preliminary report. *Quarterly Bulletin of the Archaeological Society of Virginia* 34:239–48.

Griffin, J. B.
- 1943 *The Fort Ancient aspect.* Anthropology Papers of the University of Michigan Museum of Anthropology no. 28. Ann Arbor.
- 1961 Some correlations of climatic and cultural change in eastern Northern American prehistory. *Annals of the New York Academy of Sciences* 256:710–17.
- 1978 Eastern United States. In *Chronologies in New World archaeology*, ed. R. E. Taylor and C. W. Meighan, 51–70. New York: Academic Press.
- 1983 The Midlands. In *Ancient Native Americans*, ed. J. D. Jennings, 243–302. San Francisco: Freeman.

Griffith, D. R.
- 1974 Ecological studies of prehistory. *Transactions of the Delaware Academy of Sciences* 5:63–81.
- 1977 *Townsend ceramics and the Late Woodland of southern Delaware.* Ann Arbor, Mich.: University Microfilms.
- 1980 Townsend ceramics and the Late Woodland of southern Delaware. *Maryland Historical Magazine* 75(1):23–41
- 1982 Prehistoric ceramics in Delaware: An overview. *Archaeology of Eastern North America* 10:46–68.
- n.d.a. Report on the Warrington site. Island Field Museum, South Bowers, Delaware. Typescript.
- n.d.b. Report on the Poplar Thicket site. Island Field Museum, South Bowers, Delaware. Typescript.

Griffith, D. R., and J. F. Custer
- 1983 Late Woodland ceramics of Delaware: Implications for the late prehistoric archaeology of northeastern North America. *Pennsylvania Archaeologist.* In press.

Gruber, J. W.
- 1969 Excavations at the Mohr site. *Yearbook of the American Philosophical Society 1968.* Philadelphia: American Philosophical Society.
- 1971 Patterning in death in a late prehistoric village in Pennsylvania. *American Antiquity* 36:64–76.

Guilday, J. E., P. W. Parmalee, and D. P. Tanner
- 1962 Aboriginal butchering techniques at the Eschleman site (36LA12), Lancaster County, Pennsylvania. *Pennsylvania Archaeologist* 32(2):59–83.

Guthe, A. K.
- 1958 *The late prehistoric occupation of southwestern New York: An interpretive analysis.* Rochester Museum of Arts and Sciences Research Records no. 11. Rochester, NY.

Hampton, R.
1977 An analysis of brass kettles from the Futer collection. North Museum, Franklin and Marshall College, Lancaster, PA. Typescript.

Harris, M.
1979 *Cultural materialism: The struggle for a science of culture.* New York: Random House.

Hartnett, C.
1973 Analysis of shell material from the Haverstick site (36LA8). North Museum, Franklin and Marshall College, Lancaster, PA. Typescript.

Hartzell, W.
1982 The Dark Moon site. *Newsletter of the Sussex County Historical Society,* April 1982.

Hatch, J. W.
1976 *Status in death: principles of ranking in Dallas Culture mortuary remains.* Ann Arbor, Mich.: University Microfilms.
1980 An analysis of ceramics from the stratified deposits at Fisher Farm. In *The Fisher Farm site: A Late Woodland hamlet in context,* ed. J. W. Hatch, 266–305. Pennsylvania State University Department of Anthropology Occasional Papers in Anthropology no. 12. University Park.

Hatch, J. W., and J. Daugirda
1980 The semi-subterranean keyhole structure at Fisher Farm—Feature 28. In *The Fisher Farm site: A Late Woodland hamlet in context,* ed. J. W. Hatch, 171–90. Pennsylvania State University Department of Anthropology Occasional Papers in Anthropology no. 12. University Park.

Hatch, J. W., and C. M. Stevenson
1980 A functional analysis of the Fisher Farm features. In *The Fisher Farm site: A Late Woodland hamlet in context,* ed. J. W. Hatch, 140–70. Pennsylvania State Univ. Department of Anthropology Occasional Papers in Anthropology no. 12. University Park.

Heckewelder, J.
1819 An account of the history, manners, and customs of the Indian nations who once inhabited Pennsylvania and neighboring states. *Transactions of the Historical and Literary Committee of the American Philosophical Society* Vol. 1. Philadelphia.

Heisey, H.
1971 An interpretation of Shenks Ferry ceramics. *Pennsylvania Archaeologist* 41(4):44–70.

Heisey, H, and J. P. Witmer
1962 Of historic Susquehannock cemeteries. *Pennsylvania Archaeologist* 32(3–4):99–130.

1964 The Shenks Ferry people: A site and some generalities. *Pennsylvania Archaeologist* 34(1):8–34.

Herbstritt, J. T.
1981 Bonnie Brook: A multicomponent aboriginal locus in west-central Pennsylvania. *Pennsylvania Archaeologist* 51(3):1–50.
1983 *Monongahela Cultures.* Paper presented at the Middle Atlantic Archaeological Conference, Rehoboth Beach, Delaware.

Heye, G. G., and G. H. Pepper
1915 *Explorations of a Munsee cemetery near Montague, New Jersey. Contributions of the Museum of the American Indian, Heye Foundation* no. 2. New York.

Holstein, H. O.
1979 Is the Middle Woodland period within the Monongahela drainage a time of population decrease or simply a problem of ceramic temper confusion? *Pennsylvania Archaeologist* 49(1–2):47–51.

Holzinger, C. H.
1970 *Nace site excavations.* Paper presented at the annual meeting of the Society for Pennsylvania Archaeology, Lancaster, PA.

Hudson, C.
1976 *The Southeastern Indians.* Knoxville: Univ. of Tennessee Press.

Hughes, R. B.
1980 A preliminary cultural and environmental overview of the prehistory of Maryland's lower Eastern Shore based on a survey of selected artifact collections from the area. Maryland Historical Trust, Annapolis. Typescript.

Hughes, R. B., and P. B. Weissman
1982 *Cultural resources assessment study for the selection of power plant siting areas in western Maryland.* Maryland Historical Trust Manuscript Series no. 25. Annapolis.

Hummer, C.
1978 The Late Woodland occupation of the Williamson site, Hunterdon County, New Jersey. *Newsletter of the Archaeological Society of New Jersey* 10:13–14.
n.d. Environment and Early Woodland adaptation in the Middle Delaware Valley. Typescript.

Hunter, W. A.
1959 The historic role of the Susquehannocks. In *Susquehannock Miscellany,* ed. J. Witthoft and W. F. Kinsey, 8–18. Harrisburg: Pennsylvania Historical and Museum Commission.
1974 A note on the Unalachtigo. In *A Delaware Indian Symposium,* ed. H. C. Kraft, 147–52. *Anthropological Series of the Pennsylvania Historical and Museum Commission* no. 4. Harrisburg.
1978 Documented subdivisions of the Delaware Indians. *Bulletin of the Archaeological Society of New Jersey* 35:20–39.

Hutchinson, H. H.
- 1955a Report on work done to date at the Mispillion River site. *Archeolog* 7(2):6–9.
- 1955b Progress report on the Mispillion site. *Archeolog* 7(3):2–3.
- 1967 A tentative closing site on the Willin site (18DO1). *Archeolog* 19(2).

Hutchinson, H. H., W. H. Callaway, and C. Bryant
- 1964 Report on the Chicone site #1 (18DO11) and Chicone site #2 (18DO10). *Archaeolog* 16(1):14–18.

Hutchinson, H. H., W. H. Callway, and D. Marine
- 1957 Report on the Mispillion site (7S-A-1). *Archeolog 9(2)*.

Jackson, R. W.
- 1954 The Sandy Hill Mound site. *Archeolog* 6(3).

Jennings, F.
- 1966 The Indian trade of the Susquehanna Valley. *Proceedings of the American Philosophical Society* 110:406–24.
- 1968 Glory, death, and transfiguration: The Susquehannock Indians in the seventeenth century. *Proceedings of the American Philosophical Society* 112:15–53.
- 1975 *The invasion of America: Indians, colonialism, and the cant of conquest.* Chapel Hill: Univ. of North Carolina Press.
- 1978 Susquehannocks. In *Handbook of North American Indians. Vol. 15, The Northeast*, ed. B. Trigger, 362–67. Washington, D.C.: Smithsonian Institution.
- 1982 Indians and frontiers in 17th-century Maryland. In *Maryland in a wider world*. ed. D. B. Quinn, 216–41. Detroit, Mich.: Wayne State Univ. Press.

Jochim, M.
- 1976 *Hunter-gatherer subsistence and settlement: A predictive model.* New York: Academic Press.

Johnson, W. C., J. B. Richardson, and A. S. Bohnert
- 1979 Archaeological site survey in northwest Pennsylvania, region IV. William Penn Memorial Museum, Harrisburg, Pennsylvania. Typescript.

Jones, R. W.
- 1931 Report of excavations at the Clemson and Book mounds. *Fifth Annual Report of the Pennsylvania Historical Commission*, 97–111. Harrisburg.

Jordan, F.
- 1880 *Remains of an aboriginal encampment at Rehoboth Beach, Delaware.* Philadelphia: Numismatic and Antiquarian Society.
- 1895 Aboriginal village sites of New Jersey, Delaware, and Maryland. *The Archaeologist* 3(4).
- 1906 *Aboriginal fishing stations on the coast of the Middle Atlantic states.* Lancaster, Penn.: New Era Publishing Co.

Kavanagh, M.
1979 *Archaeological reconnaissance of proposed channel improvements in the Upper Chester watershed, Kent and Queen Anne counties, Maryland.* Maryland Geological Survey, Division of Archaeology File Report no. 147. Baltimore.
1981 Prehistoric ceramics of the Monocacy Valley. Paper presented at the St. Marys City symposium on prehistoric ceramics in Maryland, St. Marys City, Maryland.
1982 *Archaeological resources of the Monocacy River region, Frederick and Carroll counties, Maryland.* Maryland Geological Survey, Division of Archaeology File Report no. 164. Baltimore.

Kent, B. C.
1974 Locust Grove pottery: A new Late Woodland variety. *Pennsylvania Archaeologist* 44(4):1–5.
1980 An update on Susquehannock Indian pottery. In *Proceedings of the 1979 Iroquois pottery conference,* ed. C. F. Hayes, 99–103. Rochester Museum and Science Center Research Records no. 13. Rochester, New York.
1984 *Susquehanna's Indians.* Harrisburg, Penn.: Pennsylvania Historical and Museum Commission.

Kent, B. C., and V. P. Packard
1969 The Erb Rockshelter. *Pennsylvania Archaeologist* 39(1–4):29–39.

Kent, B. C., I. F. Smith, and C. McCann
1971 *Foundations of Pennsylvania prehistory.* Anthropology Series of the Pennsylvania Historical and Museum Commission no. 1. Harrisburg.

Kent, B. C., J. Rice, and K. Ota
1981 A map of 18th-century Indian towns in Pennsylvania. *Pennsylvania Archaeologist* 51(4):1–18.

Kent, Brett
n.d. *Making old oysters talk: Techniques of oyster shell analysis.* Maryland Historical Manuscript Series. Annapolis. In press.

Kier, C., and F. Calverly
1957 The Raccoon Point site: An early hunting and fishing station in the lower Delaware River Valley. *Pennsylvania Archaeologist* 27(2).

Kinsey, W. F.
1957 A Susquehannock longhouse. *American Antiquity* 23:180–181.
1959 Historic Susquehannock pottery. In *Susquehannock miscellany,* ed. J. Witthoft and W. F. Kinsey, 61–98. Harrisburg: Pennsylvania Historical and Museum Commission.
1960 Additional notes on the Albert Ibaugh site. *Pennsylvania Archaeologist* 30(3–4):81–105.
1972 *Archaeology in the Upper Delaware Valley.* Anthropological Series of the Pennsylvania Historical and Museum Commission no. 2. Harrisburg.

1975	Faucett and Byram sites: Chronology and settlement in the Delaware River Valley. *Pennsylvania Archaeologist* 45(1–2).
1977	Patterning in the Piedmont Archaic: A Preliminary view. *Annals of the New York Academy of Sciences* 288:375–91.

Kinsey, W. F., and J. F. Custer

1982	Lancaster County Park site (36LA96): Conestoga phase. *Pennsylvania Archaeologist* 52(3–4):25–56.

Kinsey, W. F., and J. R. Graybill

1971	Murry site and its role in Lancaster and Funk phases of Shenks Ferry culture. *Pennsylvania Archaeologist* 41(4):7–43.

Kraft, H. C.

1965	The first petroglyph in New Jersey. *Pennsylvania Archaeologist* 35(2):93–100.
1966	Teshoas and elongated pebble tools. *Bulletin of the Archaeological Society of New Jersey* 23:1–6.
1969	There are petroglyphs in New Jersey. *Bulletin of the Archaeological Society of New Jersey* 24:13–16.
1970a	*The Miller Field site, Warren County, New Jersey: Part 1, the Archaic and Transitional stages.* South Orange, N.J.: Seton Hall Univ. Press.
1970b	Prehistoric Indian house patterns in New Jersey. *Bulletin of the Archaeological Society of New Jersey* 26:1–11.
1972	Archaeological evidence for a possible masking complex among the prehistoric Lenape in northwestern New Jersey. *Bulletin of the New York State Archaeological Association* 56:1–11.
1974a	Indian prehistory in New Jersey. In *A Delaware Indian symposium*, ed. H. C. Kraft, 1–55. Anthropological Series of the Pennsylvania Historical and Museum Commission no. 4. Harrisburg.
1974b	Feast of the dead. *Bulletin of the Archaeological Society of New Jersey* 31:29.
1974c	A petroglyph knife. *Bulletin of the Archaeological Society of New Jersey* 30:33.
1975a	*The archaeology of the Tocks Island area.* Archaeology Research Center, Seton Hall University, South Orange, New Jersey.
1975b	The Late Woodland pottery of the Upper Delaware Valley: A survey and reevaluation. *Archaeology of Eastern North America* 3:101–40.
1975c	Upside-down pendants. *Bulletin of the Archaeological Society of New Jersey* 32:33.
1976a	The archaeology of the Pahaquarra site: A preliminary report. Archaeological Research Center, Seton Hall University, South Orange, New Jersey. Typescript.
1976b	The Rosencrans site, an Adena-related mortuary complex in the Upper Delaware Valley, New Jersey. *Archaeology of Eastern North America* 4:9–50.

1977 *The Minisink settlements: An investigation into a prehistoric and early historic site in Sussex County, New Jersey.* Archaeological Research Center, Seton Hall University, South Orange, New Jersey.

1978 *The Minisink site: A reevaluation of a late prehistoric and early historic contact site in Sussex County, New Jersey.* Archaeological Research Center, Seton Hall University, South Orange, New Jersey.

1980 Effigy-bearing celt. *Bulletin of the Archaeological Society of New Jersey* 36:38.

Kraft, H., and A. Mounier
1982 The Late Woodland period in New Jersey. In *New Jersey's archaeological resources from the Paleo-Indian period to the present: A review of research problems and survey priorities,* ed. O. Chesler, 139–84. Trenton: Office of Cultural and Environmental Services, New Jersey Department of Environmental Protection.

Kraft, J. C., and C. John
1978 Paleogeographic analysis of coastal archaeological settings in Delaware. *Archaeology of Eastern North America* 6:41–59.

Kristiansen, K.
1982 The formation of tribal systems in later European prehistory: Northern Europe, 4000–500 B.C. In *Theory and explanation in archaeology,* ed. C. Renfrew, M. J. Rowlands, and B. A. Segraves, 241–80. New York: Academic Press.

Kroeber, A. L.
1955 The nature of the land-holding group. *Ethnohistory* 2:303–14.
1963 The nature of land-holding groups in California. In *Aboriginal California,* ed. R. F. Heizer, 81–120. Berkeley and Los Angeles: Univ. of California Press.

Kurtzman, E.
1974 A preliminary analysis of faunal remains from Wood Rockshelter. North Museum, Franklin and Marshall College, Lancaster, Pennsylvania. Typescript.

Laet, J. de
1909 "From the New World," by Johan de Laet, 1625, 1630, 1633, 1640. In *Narratives of New Netherland 1609–1664,* ed. J. Franklin Jameson, 36–60. New York: Charles Scribners.

Landis, S.
1982 *Archaeological investigations of the Point Pleasant area: A preliminary report.* Doylestown, Penn. Bucks County Conservancy.

Larson, H. H.
1971 Archaeological implications of social stratification at the Etowah site, Georgia. In *Approaches to the social dimensions of mortuary*

practice, Memoirs of the Society for American Archaeology no. 25, ed. J. A. Brown, 58–67. Washington, D.C.: Society for American Archaeology.

Leidy, J.
1865 Report of investigations. *Proceedings of the Philadelphia Academy of Sciences,* June 1865–October 1866.

Lindestrom, P.
1925 *Geographia Americae with an account of the Delaware Indians based on surveys and notes made in 1654–1656 (1691),* ed. A. Johnson. Philadelphia: The Swedish Colonial Society.

Lopez, J.
1961 Pottery from the Mispillion site, Sussex County, Delaware, and related types in surrounding areas. *Pennsylvania Archaeologist* 31(1):1–38.

Lorant, S.
1946 *The New World, the first pictures of America.* New York: Dvell, Sloan, and Pearce.

Louis Berger and Associates
1982a *Phase II cultural resource surveys and mitigation plans in the Trenton vicinity.* Federal Highway Administration and the New Jersey Department of Transportation, Trenton.
1982b *Proposal for data recovery at the Shady Brook site, 28ME20 and 28ME99, Hamilton Township, Mercer County, New Jersey, Phase III mitigation.* Federal Highway Administration and the New Jersey Department of Transportation, Trenton.
1982c *Abbott Farm National Landmark: Phase II cultural resource survey and mitigation plan.* Federal Highway Administration and the New Jersey Department of Transportation, Trenton.

Lucy, C. E.
1959 Pottery types of the Upper Susquehanna. *Pennsylvania Archaeologist* 29(1):28–37.

McCann, C.
1971 Notes on the pottery of the Clemson and Book mounds. In *Foundations of Pennsylvania prehistory,* ed. B. Kent, 419–23. Anthropological Series of the Pennsylvania Historical and Museum Commission no. 1. Harrisburg.

McCary, B. C.
1950 A burial site, Richmond County, Virginia. *Quarterly Bulletin of the Archaeological Society of Virginia 5(1).*
1953 The Potts site, Chicahominy River, New Kent County, Virginia. *Quarterly Bulletin of the Archaeological Society of Virginia* 8(1).
1958a A conch shell mask found in Virginia. *Quarterly Bulletin of the Archaeological Society of Virginia* 12(4).
1958b Further notes on the Melton mask. *Quarterly Bulletin of the Archaeological Society of Virginia* 13(2).

1958c   The Kiskiack (Chiskiack) Indian site near Yorktown, Virginia. *Quarterly Bulletin of the Archaeological Society of Virginia* 13(2).
1967   The Briarfield site. *Quarterly Bulletin of the Archaeological Society of Virginia* 2:67–74.

McCary, B. C., and N. F. Barka
1977   The John Smith and Zuniga maps in light of recent archaeological investigations along the Chickahominy River. *Archaeology of Eastern North America* 5:73–86.

MacCord, H. A.
1964a   The Irwin site, Prince George County, Virginia. *Quarterly Bulletin of the Archaeological Society of Virginia* 19:37–42.
1964b   The Phillip Nase site, Henrico County, Virginia. *Quarterly Bulletin of the Archaeological Society of Virginia* 18:78–85.
1965   The DeShazo site, King George County, Virginia. *Quarterly Bulletin of the Archaeological Society of Virginia* 19:98–104.
1967   The Hopewell Airport site, Prince George County, Virginia. *Quarterly Bulletin of the Archaeological Society of Virginia* 22:73–80.
1969   Camden: A postcontact Indian site in Caroline County. *Quarterly Bulletin of the Archaeological Society of Virginia* 24:1–55.
1974   The Refo site, Mathews County, Virginia. *Quarterly Bulletin of the Archaeological Society of Virginia* 29:33–39.
1975   The Kiser site, Chesterfield County, Virginia. *Proceedings of the Eastern States Archaeological Federation* 34:16.

MacCord, H. A., and R. M. Owens
1965   The Posnick site, Henrico County, Virginia (44HE3). *Quarterly Bulletin of the Archaeological Society of Virginia* 19:88–96.

MacCord, H. A., K. Schmitt, and R. G. Slattery
1957   The Shepard site study. *Bulletin of the Archaeological Society of Maryland* no. 1. Baltimore.

McNamara, J. M.
1982a   *Summary of 1981 excavations at the Conowingo site, 18CE14.* Maryland Geological Survey, Division of Archaeology File Report No. 172. Baltimore.
1982b   *Archaeological test excavations at the Hillsmere Pond I site, Anne Arundel County, Maryland.* Division of Archaeology, Maryland Geological Survey Report no. 157. Baltimore.
1983a   *Summary of 1982 excavations at the Conowingo site, 18CE14.* Maryland Geological Survey, Division of Archaeology File Report no. 176. Baltimore.
1983b   The effects of mid-postglacial changes at Conowingo: A stratified Lake Archaic-Late Woodland site in the Piedmont floodplain of the Lower Susquehanna Valley. Paper presented at the Middle Atlantic Archaeological Conference, Rehoboth Beach, Delaware.

McNett, C., and W. M. Gardner
- n.d. Archaeology in the Middle and Lower Potomac Valley. Department of Anthropology, American University, Washington, D.C. Typescript.

Mann, R. W.
- 1981 Edgehill site: Analysis and interpretations of a Virginia ossuary. Senior thesis, College of William and Mary, Williamsburg.

Manson, C., H. A. MacCord, and J. B. Griffin
- 1944 The culture of the Keyser Farm site. *Papers of the Michigan Academy of Sciences, Arts, and Letters* 29:375–418.

Mansueti, R. J., and H. Kolb
- 1953 A historical review of the shad fisheries of North America. Chesapeake Biological Laboratory Publications No. 97. Solomon, Maryland.

Marine, D.
- 1957 Report on the Russell site. *Archeolog* 9(1):1–9.

Marine, D., J. L. Parsons, and K. Hall
- 1965 Report on an outlying shell midden of the Rehoboth City site. *Archeolog* 17(1):18–24.

Marine, D., M. Tull, F. Austin, J. Parsons, and H. Hutchinson
- 1964 Report on the Warrington site (7S-G-14). *Archeolog* 16(1):1–13.

Marshall, B.
- 1977 Report on an intensive archaeological reconnaissance survey of Pine Bluff Village, Wicomico County, Maryland (18WC16 and 18WC20). Division of Archaeology, Maryland Geological Survey, Baltimore. Typescript.

Martin, K.
- 1974 The foraging adaptation: Uniformity or diversity? *Addison-Wesley Module in Anthropology* No. 56. New York: Addison-Wesley.

Marye, W. B.
- 1936a Indian paths of the Delmarva Peninsula. *Bulletin of the Archaeological Society of Delaware* 2(3):5–22.
- 1936b Indian paths of the Delmarva Peninsula, Part I. *Bulletin of the Archaeological Society of Delaware* 2(4):5–27.
- 1937 Indian paths on the Delmarva Peninsula, Part II. *Bulletin of the Archaeological Society of Delaware* 2(5):1–37.
- 1938 Indian paths of the Delmarva Peninsula, Part III. *Bulletin of the Archaeological Society of Delaware* 2(6):4–11.
- 1939 Indian towns of the southwestern part of Sussex County. *Bulletin of the Archaeological Society of Delaware* 3(2):18–25.
- 1940 Indian towns of the southwestern part of Sussex County. *Bulletin of the Archaeological Society of Delaware* 3(3):21–28.

Maryland Geological Survey, Division of Archaeology, Johns Hopkins University, Baltimore.

Maryland Historical Trust
- 1974a National Register nomination for the Buckingham archaeological site. Division of Archaeology, Maryland Geological Survey, Baltimore. Typescript.
- 1974b National Register nomination for the Brinsfield I prehistoric archaeological site. Division of Archaeology, Maryland Geological Survey, Baltimore. Typescript.
- 1974c National Register nomination for the Willin Village archaeological site. Division of Archaeology, Maryland Geological Survey, Baltimore, Typescript.
- 1974d National Register nomination for the Sandy Point archaeological site. Division of Archaeology, Maryland Geological Survey, Baltimore. Typescript.

Mayer-Oakes, W. J.
- 1955 *Prehistory of the Upper Ohio Valley.* Vol. 34 of Annals of the Carnegie Museum. Pittsburgh, Pennsylvania.

Mercer, H.
- 1897 *The antiquity of man in the Delaware Valley and the Eastern United States.* Publications of the University of Pennsylvania Series on Philology, Literature, and Archaeology Vol. 6. Philadelphia.

MGS site files. *See* Maryland Geological Survey.

MHT. *See* Maryland Historical Trust.

Michels, J. W.
- 1967 A culture history of the Sheep Rock Shelter. In *Archaeological Investigations of Sheep Rock Shelter, Huntington County, Pennsylvania, Vol. II*, ed. J. W. Michels and I. F. Smith, 801–24. University Park: Pennsylvania State Univ. Department of Anthropology.

Milisauskas, S.
- 1978 *European prehistory.* New York: Academic Press.

Miller, J. P., J. W. Friedusdorff, and H. Mears
- 1974 *Annual progress report of the Delaware River basin anadramous fish project.* Rosemont, N.J.: U.S. Fish and Wildlife Service and the Pennsylvania Fish Commission.

Moeller, R. W.
- 1975 Late Woodland floral and faunal exploitative patterns in the Upper Delaware Valley. In *Proceedings of the 1975 Middle Atlantic Archaeological Conference*, ed. W. F. Kinsey, 51–56. Lancaster, Penn.: North Museum, Franklin and Marshall College.

Mouer, L. D.
- 1981 Powhatan and Monacan regional settlement hierarchies: A model of relationships between social and environmental structure.

Quarterly Bulletin of the Archaeological Society of Virginia 36:1–21.

Mounier, R. A.
1975 The Indian Head site revisited. *Bulletin of the Archaeological Society of New Jersey* 32:1–14.
1979 *An archaeological survey of Interstate Highway 295: Sections IX to IW.* New Jersey Department of Transportation, Trenton.

Muller, J.
1983 The Southeast. In *Ancient North Americans*, J. D. Jennings, 373–421. San Francisco: Freeman.

Murray, P.
1980 Discard location: The ethnographic data. *American Antiquity* 45:490–501.

Naroll, R.
1956 A preliminary index of social development. *American Anthropologist* 58:687–716.

Neill, E. D.
1968 *History of the Virginia Company of London with letters to and from the first colony never before printed.* New York: Burt Franklin.

NJMR (New Jersey Museum Report)
1905 *The fishes of New Jersey.* Annual report of the New Jersey State Museum. Trenton.

Newcomb, W.
1956 *The culture and acculturation of the Delaware Indians.* Anthropological Papers of the University of Michigan Museum of Anthropology no. 10. Ann Arbor.

Oberg, K.
1955 Types of social structure among lowland tribes of South and Central America. *American Anthropologist* 57:472–87.

Omwake, G.
1945 Refuse pits on Sinepuxent Neck on the Eastern Shore of Maryland. *Bulletin of the Archaeological Society of Delaware* 4(2):2–11.
1948 For the record. *Archeolog* 1(2):6–10.
1951 Preliminary comments on the Ritter site near Lewes, Delaware. *Archeolog* 3(2):7–8.
1954a Notes on the Phillips-Robinson-Benson site near Milford Delaware. *Archeolog* 6(1):1–2.
1954b A report on the excavations at the Ritter no. 2 site near Lewes, Delaware. *Archeolog* 6(3):4–12.
1954c A report on the Miller-Toms site (7S-D-4). *Archeolog* 6(2):3–10.
1954d A report on the archaeological investigation of the Ritter site, Lewes, Delaware. *Archeolog 6(1):24–39.*

Omwake, G., and T. D. Stewart
1963   The Townsend site near Lewes, Delaware. *Archeolog* 15(1):1–72.

Owens, R. M.
1969   Martin Farm, New Kent County. *Quarterly Bulletin of the Archaeological Society of Virginia* 24:81–116.

Painter, F.
1980   The Great King of Great Neck: A status burial from Coastal Virginia. *Chesopiean* 18(3–6):75–79.

Parris, D.
1980   Faunal evidence of seasonal occupation of the Abbott Farm locality. Paper presented at the annual meeting of the Society for American Archaeology, Philadelphia.

Parry, W. J.
1975   The prehistoric human ecology of Washington Boro, Lancaster County, Pennsylvania. *The Kithuwan, Journal of the Anthropology Club, Franklin and Marshall College* 7:6–19.

Payne, W.
1983   Site report, prehistoric aboriginal site, Holiday Park. University of Delaware Center for Archaeological Research, Newark. Typescript.

Peck, D.
1979   Archaeological resources of the Monocacy River region. Maryland Geological Survey, Division of Archaeology, Baltimore. Typescript.

Peck, D., and T. Bastian
1977   Test excavations at the Devilbiss site, Frederick County, Maryland. *Maryland Archaeology* 13(2):1–10

Peebles, C. S.
1971   Moundville and surrounding sites: Some structural considerations of mortuary practices II. In *Approaches to the social dimensions of mortuary practices*, ed. J. A. Brown, 68–91. Memoirs of the Society for American Archaeology no. 25.

Peebles, C. S., and S. M. Kus
1977   Some archaeological correlates of ranked societies. *American Antiquity* 42:421–48.

Peebles, P. W.
1983   Hatch site progress report. *Newsletter of the Archaeological Society of Virginia* 84:1–3.

Philhower, C. A.
1953   The historic Minisink site, part 1. *Bulletin of the Archaeological Society of New Jersey* 7:1–9.

Plog, S.
1980   *Stylistic variation in prehistoric ceramics: Design analysis in the American Southwest.* New York: Cambridge Univ. Press.

Pollak, J.
- 1968 Salvage excavation at the Watson House, Abbott Farm site. *Bulletin of the New Jersey Academy of Science* 13:84.
- 1975 Field diary for New Jersey Department of Transportation survey. New Jersey State Museum, Trenton. Typescript.
- 1977 Prehistoric resources. In *Case report, cities of Trenton and Bordentown, townships of Hamilton and Bordentown, Route I-195, . . .Counties of Mercer and Burlington, State of New Jersey*. Federal Highway Administration and the New Jersey Department of Transportation, Trenton.

Porter, F. W.
- 1979 *Indians in Maryland and Delaware: A Critical bibliography*. Bloomington: Indiana Univ. Press.

Potter, S.
- 1980 *An overview of the archaeological resources of Piscataway Park*. Washington, D.C.: National Park Service.
- 1982 An analysis of Chicacoan settlement patterns. Ph.D. diss., University of North Carolina, Chapel Hill.

Price, B.
- 1981 Productive intensification and ranked society: speculations from evolutionary theory. Department of Anthropology, Columbia University, New York. Typescript.
- 1982 Cultural materialism: A theoretical overview. *American Antiquity* 47:709–41.

Puniello, A. J.
- 1980 Iroquois series ceramics in the Upper Delaware Valley, New Jersey, and Pennsylvania. In *Proceedings of the 1979 Iroquois pottery conference* ed. C. F. Hayes, 146–55. Rochester Museum and Science Center Research Records no. 13. Rochester, New York.

Purnell, H. W. T.
- 1958 The Draper site. *Archeolog* 10(2):1–16.

Quinn, D. B., ed.
- 1955 *The Roanoke voyages*. London: Cambridge Univ. Press.

Rainey, F. G.
- 1956 A compilation of historical data contributing to the ethnography of Connecticut and southern New England Indians. *Bulletin of the Archaeological Society of Connecticut* 3:3–49.

Reinhart, T. R.
- 1978 Plow zone archeology on College Creek, James City County, Virginia. *Quarterly Bulletin of the Archaeological Society of Virginia* 32:81–93.

Renfrew, C.
- 1974 Beyond a subsistence economy: The evolution of social organization in prehistoric Europe. *Bulletin of the American School of Oriental Research, Supplement* 20:69–88.

Reynolds, E. R.
1883   *Ossuary of the Accotink, Virginia. Abstract of Transactions of the Anthropological Society of Washington (1881).* Smithsonian Miscellaneous Collections 25. Washington, D.C.

Richardson, A. S.
1884   A. S. Richardson calls attention to the existence of two mounds in the vicinity of West Point, King William County, Virginia. *Annual Report of the Smithsonian Institution for the year 1882.* Washington, D.C.

Ritchie, W. A.
1949   *The Bell-Philhower site, Sussex County, New Jersey.* Indiana Historical Society, Prehistoric Research Series 3(2). Indianapolis.
1965   *The archaeology of New York State.* Garden City, N.Y.: Natural History Press.
1969   *The archaeology of New York State,* rev. ed. Garden City, N.Y.: Natural History Press.

Ritchie, W. A., and R. E. Funk
1973   *Aboriginal settlement patterns in the Northeast.* New York State Museum and Science Service Memoir 20. Albany.

Robertson, L. B., and B. P. Robertson
1978   *The New River survey: A preliminary report.* Archaeology Branch, Division of Archives and History, North Carolina Department of Cultural Resources, Raleigh.

Rothschild, N. A.
1979   Mortuary behavior and social organization at Indian Knoll and Dickson Mounds. *American Antiquity* 44:658–73.

Sahlins, M. D.
1958   *Social stratification in Polynesia.* Seattle: Univ. of Washington Press.
1961   The segmentary lineage: An organization of predatory expansion. *American Anthropologist* 63:322–45.
1968   *Tribesmen.* Englewood Cliffs, N.J.: Prentice-Hall.

Sahlins, M. D., and E. Service
1960   *Evolution and culture.* Ann Arbor: Univ. of Michigan Press.

Sanders, W. T., and J. Marino
1970   *New World Prehistory.* Englewood Cliffs, N. J.: Prentice-Hall.

Sanders, W. T., and B. J. Price
1968   *Mesoamerica: The evolution of a civilization.* New York: Random House.

Sanders, W. T., and D. Webster
1978   Unilinealism, multilinealism, and the evolution of complex societies. In *Social Archaeology: Beyond subsistence and dating,* ed. C. L. Redman, 249–302. New York: Academic Press.

Schmitt, K.
1952   Archaeological chronology of the Middle Atlantic states. In *Archaeology of Eastern United States*, ed. J. B. Griffin, 59–70. Chicago: Univ. of Chicago Press.
1965   Patawomeke: An historic Algonkian site. *Quarterly Bulletin of the Archaeological Society of Virginia* 20:1–36.

Schrabish, M.
1915   Indian habitations in Sussex County, New Jersey. *Bulletin of the Geological Survey of New Jersey* 13. Trenton.

Seeman, M. F.
1979   Feasting with the dead: Ohio charnel house ritual as a context for redistribution. In *Hopewell archaeology, the Chillicothe conference*, ed. D. S. Brose and N. Greber, 39–46. Kent, Ohio: Kent State Univ. Press.

Service, E.
1962   *Primitive social organization*. New York: Random House.
1968   *The hunters*. Englewood Cliffs, N. J.: Prentice-Hall.

Shepard, A. O.
1954   *Ceramics for the archaeologist*. Washington, D. C.: Carnegie Institute.

Skinner, A.
1914   Report on the Trenton simple culture. Anthropology Department, the American Museum of Natural History, New York. Typescript.
1915   Archaeological research in New Jersey. Anthropology Department, the American Museum of Natural History, New York. Typescript.

Slattery, R. G., W. A. Tidwell, and R. G. Woodward
1966   The Montgomery focus. *Quarterly Bulletin of the Archaeological Society of Virginia* 21:49–51.

Smith, B., ed.
1978   *Mississippian settlement patterns*. New York: Academic Press.

Smith, C.
1950   The archaeology of coastal New York. *Anthropological Papers of the American Museum of Natural History* 43(2). New York.
1984   *A Late Woodland village site in north central Pennsylvania*. Harrisburg, Penn.: William Penn Memorial Museum.

Smith, I. F.
1973   The Parker site: A manifestation of the Wyoming Valley culture. *Pennsylvania Archaeologist* 43(3–4):1–56.
1976   A functional analysis of keyhole structures in the Northeast. *Pennsylvania Archaeologist* 46(1–2):1–12.
1978   *A description and analysis of early pottery types in the lower Susquehanna Valley of Pennsylvania*. Harrisburg: Pennsylvania Historical and Museum Commission.

Smith, J.
1965 The description of Virginia (1607). In *Hakluytus posthumus or Purchas his pilgrimes,* by S. Purchas, 420–58. New York: AMS Press.

Smith, S.
1890 *The history of the colony of Nova-Caesaria, or New Jersey.* Trenton, N. J.: William S. Sharp.

Snow, D. R.
1980 *The archaeology of New England.* New York: Academic Press.

Snyder, D.
1975 The Kibler-Funk site (36LA205). North Museum, Franklin and Marshall College, Lancaster, Pennsylvania. Typescript.

Speck, F.
1927 *The Nanticoke and Conoy Indians with a review of linguistic material from manuscript and living sources: An historical account.* Papers of the Historical Society of Delaware n.s. no. 1. Wilmington.
1928 *Chapters on the ethnology of the Powhatan tribes of Virginia.* Indian Notes and Monographs 1 (5). Museum of the American Indian, Heye Foundation, New York.

Spier, L.
1918 *The Trenton argillite culture.* Anthropological Papers of the American Museum of Natural History 22(4). New York.

Staats, F. D.
1974 A fresh look at Bowmans Brook and Overpeck incised pottery. *Bulletin of the Archaeological Society of New Jersey* 30:1–7.
1977 The Owasco corded-collar and related pottery types of the Upper Delaware Valley. *Bulletin of the Archaeological Society of New Jersey* 34:1–7.

Stanzeski, A.
1981 Abbott Farm project, survey of individual collectors. Louis Berger and Associates, East Orange, New Jersey. Typescript.

Stearns, R.
1943 *Some Indian village sites of Tidewater Maryland.* Proceedings of the Natural History Society of Maryland no. 9. Baltimore.

Steffy, D. M.
1968 A preliminary report on the investigation of subsistence-oriented plant utilization at the Sheep Rock Shelter site. In *Archaeological Investigations of the Sheep Rock Shelter, Huntingdon County, Pennsylvania, Vol. III.* ed. J. W. Michels and S. Dutt, 165–80. University Park: Pennsylvania State University Department of Anthropology.

Stephenson, R.
- 1963 *The Accokeek Creek site: A Middle Atlantic seaboard culture sequence.* Anthropological Papers of the University of Michigan Museum of Anthropology no. 20. Ann Arbor.

Steponaitis, L. C.
- 1980 *A survey of artifact collections from the Patuxent River drainage, Maryland.* Maryland Historical Trust Monograph Series no. 1. Annapolis.
- 1983 An archaeological survey of the Patuxent River drainage. Paper presented at the Middle Atlantic Archaeological Conference, Rehoboth Beach, Delaware.

Steponaitis, V.
- 1978 Location theory and complex chiefdoms: A Mississippian example. In *Mississippian settlement patterns,* ed. B. Smith, 417–53. New York: Academic Press.

Stevenson, C.
- 1982 Patterns of hollow exploitation along the Allegheny Front, Centre County, Pennsylvania. *Pennsylvania Archaeologist* 52(3-4):1–16.

Steward, J. H.
- 1948 The circum-Caribbean tribes: An introduction. *Bulletin of the Bureau of American Ethnology* 143:1–41.
- 1955 *Theory of culture change: The methodology of multilinear evolution.* Urbana: Univ. of Illinois Press.

Steward, J. H., and L. C. Faron
- 1959 *Native peoples of South America.* New York: McGraw-Hill.

Stewart, R. M.
- 1980 *Prehistoric settlement-subsistence patterns and the testing of predictive site location models in the Great Valley of Maryland.* Ann Arbor, Mich.: University Microfilms.
- 1981a Prehistoric burial mounds in the Great Valley of Maryland. *Maryland Archaeology* 17(1):1–16.
- 1981b *The Shady Brook site (28ME20, I-295, Arena Drive interchange: Report of the archaeological investigations for the determination of eligibility to the National Register of Historic Places.* Federal Highway Administration and the New Jersey Department of Transportation, Trenton.
- 1982a The Middle Woodland of the Abbott Farm: Summary and hypotheses. In *Practicing environmental archaeology: Methods and interpretations,* ed. R. Moeller, 19–28. Occasional Papers of the American Indian Archaeological Institute no. 3, Washington, Connecticut.

1982b　The Late Woodland of the Abbott Farm. Paper presented at the annual meeting of the Society of Pennsylvania Archaeology, Harrisburg.

1982c　Rethinking the Abbott Farm: Oral tradition, context, and historical perspective. Paper presented at the annual meeting of the Eastern States Archaeological Federation, Norfolk, Virginia.

1982d　Prehistoric ceramics of the Great Valley of Maryland. *Archaeology of Eastern North America* 10.

1983　Prehistoric settlement patterns in the Blue Ridge province of Maryland. In *Upland archaeology in the East: A symposium*, ed. C. R. Geier, M. B. Barber, G. A. Tolley, 43–90. Washington, D.C.: U.S. Forest Service.

1984　Excavations at Shady Brook. New Jersey Department of Transportation, Trenton. Typescript.

n.d.　*Phase III mitigation of the Shady Brook site (28ME20)*. Federal Highway Administration and the New Jersey Department of Transportation, Trenton.

Stewart, R. M., and W. M. Gardner
1978　*Phase II archaeological investigations near Sam Rice Manor, Montgomery County, and at 18PR166 and 18PR172 near Accokeek, Prince Georges County, Maryland*. Washington Suburban Sanitary Commission, Washington, D.C.

Stewart, T. D.
1939　Excavating the Indian village of Patawomeke (Potomac). *Explorations and Fieldwork of the Smithsonian Institution in 1938*, Washington, D.C.

1940a　Further excavations at the Indian village site of Patawomeke (Potomac). *Explorations and Fieldwork of the Smithsonian Institution in 1939*, Washington D.C.

1940b　The finding of an Indian ossuary on the York River in Virginia. *Journal of the Washington Academy of Sciences* 30:356–64.

1941　An ossuary at the Indian village of Patawomeke (Potomac). *Explorations and Fieldwork of the Smithsonian Institution in 1940*, Washington, D.C.

1945　Skeletal remains from the Rehoboth Bay ossuary. *Bulletin of the Archaeological Society of Delaware* 4(2):24–25.

Stocum, F.
1977　An important ceramic discovery at the Robbins Farm site. *Bulletin of the Archaeological Society of Delaware* Fall 1977:40–48.

Strachey, W.
1953　*The history of travell into Virginia Britania*. London: Cambridge Univ. Press.

Strohmeier, W.
1980　The Unami Creek rockshelter. *Pennsylvania Archaeologist* 50(4):1–12.

Struthers, T., and D. Roberts
1983   *The Lambertville site (28HU468: An Early-Middle and Late Woodland site in the Middle Delaware River Valley.* Lambertville Sewage Authority and the Environmental Protection Agency, Region 2; Lambertville, New Jersey.

Swanton, J. R.
1946   *The Indians of the Southeastern United States.* Bulletin of the Bureau of American Ethnology no. 137. Washington, D.C.

Swientochowski, J., and C. A. Weslager
1942   Excavations at the Crane Hook site, Wilmington, Delaware. *Bulletin of the Archaeological Society of Delaware* 3(5):2–17.

Sykes, C. M.
1980   Swidden horticulture and Iroquoian settlement. *Archaeology of Eastern North America* 8:45–52.

Sykes, J. E., and B. Q. Lehman
1957   *Past and present Delaware River shad fishery and considerations for its future.* United States Fish and Wildlife Service Research Report 46. Washington, D.C.

Taylor, D.
1975   Some locational aspects of middle-range hierarchical societies. Ph.D. diss., City College of New York. University Microfilms, Ann Arbor, Michigan.

Thomas R. A.
1966a   7NC-F-7, The Hell Island site. *Delaware Archaeology* 2(2):1–18.
1966b   Archaeological investigations on Milford Neck. *Delaware Archaeology* 2(4)
1973   Prehistoric mortuary complexes of the Delmarva Peninsula. In *Proceedings of the 1973 Middle Atlantic Archaeological Conference,* ed. R. A. Thomas, 50–72. Penns Grove, N.J.: Middle Atlantic Archaeological Research, Inc.
1974   Webb phase mortuary customs at the Island Field site. *Transactions of the Delaware Academy of Sciences* 5:49–61.
1977   Radiocarbon dates of the Woodland period from the Delmarva Peninsula. *Bulletin of the Archaeological Society of Delaware* 11:49–57.
1978   A cultural resources reconnaissance for the Wicomico River (east) federal maintenance dredging project. Division of Archaeology, Maryland Geologial Survey, Baltimore. Typescript.
1981   *Excavations at the Delaware Park site (7NC-E-41)* Dover: Delaware Department of Transportation.
1982   Intensive archaeological investigations at the Hollingsworth Farm site, Elkton, Maryland. *Maryland Archaeology* 18(1):9–28.

Thomas R. A., D. R. Griffith, C. L. Wise, and R. E. Artusy
1975   Environmental adaptation on Delaware's coastal plain. *Archaeology of Eastern North America* 3:35–90.

Thomas, R. A., and N. Warren
1970a A Middle Woodland cemetery in central Delaware: Excavations at the Island Field site. *Bulletin of the Archaeological Society of Delaware* 8.
1970b Salvage excavations of the Mispillion site. *Archeolog* 22(2):1–23.

Thompson, T. A.
1982 Investigations at 18KE246, Eastern Neck National Wildlife Refuge, Kent County, Maryland. Maryland Historical Trust, Annapolis. Typescript.

Thompson, T. A., and W. M. Gardner
1978 A cultural resources reconnaissance and impact areas assessment of the Eastern Neck Wildlife Refuge, Kent County, Maryland. Department of Anthropology, Catholic University of America, Washington, D.C. Typescript.

Thurman, M. D.
1974 Delaware social organization. In *Delaware Indian symposium*, ed. H. C. Kraft, 111–34. Anthropological Series of the Pennsylvania Historical and Museum Commission no. 4. Harrisburg.

Tirpak, R.
1978 Activity analysis: A technique for the possible determination of seasonal occupation at the Mispillion site. *Bulletin of the Archaeological Society of Delaware* 11:22–68.

Tolley, G. A.
1983 Blue Ridge prehistory: Perspective from the George Washington National Forest. In *Upland archaeology in the East: A symposium*, ed. C. R. Geier, M. B. Barber, and G. A. Tolley, 104–15. Washington, D.C.: U.S. Forest Service.

Trigger, B.
1978 Iroquois matriliny. *Pennsylvania Archaeologist* 48(1–2):55–65.
1981 Prehistoric social and political organization: An Iroquoian case study. In *Foundations of northeastern archaeology*, ed. D. R. Snow, 1–50. New York: Academic Press.

Tuck, J. A.
1978 Northern Iroquoian prehistory. In *Handbook of North American Indians. Vol. 15, The Northeast*, ed. B. Trigger, 322–33. Washington, D.C.: Smithsonian Institution.

Turnbaugh, W. A.
1977 *Man, land, and time.* Evansville, Penn.: Unigraphic.

Turner, E. R.
1976 *An archaeological and ethnohistorical study on the evolution of rank societies in the Virginia Coastal Plain.* Ann Arbor, Mich.: University Microfilms.
1981 The archaeological identification of chiefdom societies in southwestern Virginia. Paper presented at the symposium, Upland

Archaeology in the East. James Madison University, Harrisonburg, Virginia.
- 1982a   Socio-political organization within the Powhatan chiefdom and the effects of European Contact, A.D. 1607–1646. Paper presented at symposium, European and Indian Adaptations along the Chesapeake Frontier. Anthropological Society of Washington, Washington, D.C.
- 1982b   A re-examination of Powhatan territorial boundaries and population ca. A.D. 1607. *Quarterly Bulletin of the Archaeological Society of Virginia* 37:45–64.

Ubelaker, D. H.
- 1974   *Reconstruction of demographic profiles from ossuary skeletal samples: A case study from the tidewater Potomac.* Smithsonian Contributions to Anthropology no. 18. Washington, D.C.

Volk, E.
- 1911   *The archaeology of the Delaware Valley.* Papers of the Peabody Museum of American Archaeology and Ethnology. Harvard University Vol. 5. Cambridge, Mass.

VRCA site files (Virginia Research Center for Archaeology). Yorktown.

Walberg, C. H., and P. R. Nichols.
- 1967   *Biology and management of the American shad and status of the Atlantic Coast of the United States, 1960.* United States Fish and Wildlife Service, Special Scientific Report—Fisheries no. 550. Washington, D.C.

Wall, R. D.
- 1981   *An archaeological study of the Maryland coal region: The prehistoric resources.* Baltimore: Maryland Geological Survey.

Wallace, P.
- 1961   *Indians in Pennsylvania.* Harrisburg: Pennsylvania Historical and Museum Commission.

Wanser, J. C.
- 1982   *A survey of artifact collections from central southern Maryland.* Maryland Historical Trust Manuscript Series no. 23. Annapolis.

Waselkov, G. A.
- 1982   Shellfish gathering and shell midden archaeology. Ph.D. diss. University of North Carolina, Chapel Hill.

Wells, I.
- 1981   A spatial analysis methodology for predicting archaeological sites in Delaware and its potential application in remote sensing. Master's thesis, College of Marine Studies, University of Delaware, Newark.

Wells, I., J. F. Custer, and V. Klemas
- 1981   Locating prehistoric sites by remote sensing using LANDSAT. *Proceedings of the 15th International Symposium on Remote Sens-*

ing of the Environment. Ann Arbor, Mich.: International Society for Remote Sensing.

Weslager, C. A.
1939   An aboriginal shell heap near Lewes, Delaware. *Bulletin of the Archaeological Society of Delaware* 3(2).
1942   Ossuaries on the Delmarva Peninsula and exotic influences in the coastal aspect of the Woodland period. *American Antiquity* 8:141–51.
1953   *Red men on the Brandywine.* Wilmington, Del.: Hamilton Co.
1954   Robert Evelyn's Indian tribes and place-names of New Albion. *Bulletin of the Archaeological Society of New Jersey* 9:1–14.
1961   *Dutch explorers, traders, and settlers in the Delaware Valley, 1609–1664.* Philadelphia: Univ. of Pennsylvania Press.
1968   *Delaware's buried past.* New Brunswick, N.J.: Rutgers Univ. Press.
1972   *The Delaware Indians: A History.* New Brunswick, N.J.: Rutgers Univ. Press.
1978   *The Delawares: A critical bibliography.* Bloomington: Indiana Press.
1983   *The Nanticoke Indians—past and present.* Newark: Univ. of Delaware Press.

Whallon, R.
1968   Investigations of late prehistoric social organization in New York state. In *New Perspectives in Archaeology,* ed. L. Binford and S. Binford, 223–44. Chicago: Aldine.

Wigglesworth, J.
1933   Excavations at Rehoboth. *Bulletin of the Archaeological Society of Delaware* 3:2–6.

Wilke, S.; and G. Thompson
1976   *Prehistoric resources of portions of coastal Kent County, Maryland.* Maryland Geological Survey, Division of Archaeology File Report no. 139. Baltimore.

Wilkins, E.
1976   The lithics of the Delaware and Nanticoke Indians. *Transactions of the Delaware Academy of Sciences* 74:25–35.

Willey, G. R., and P. Phillips
1958   *Method and theory in American archaeology.* Chicago: Univ. of Chicago Press.

Willey, L. M.
1980   The analysis of flotation samples from the Fisher Farm site. In *The Fisher Farm site: A Late Woodland hamlet in context,* ed. J. W. Hatch, 136–39. Pennsylvania State University Department of Anthropology Occasional Papers in Anthropology no. 12. University Park.

Williams, L., D. Parris, and S. Albright
- 1981 Interdisciplinary approaches to WPA archaeological collections in the Northeast. Paper presented at the annual meeting of the New York Academy of Sciences, New York.

Willliams, L., and R. Thomas
- 1982 The Early/Middle Woodland in New Jersey. In *New Jersey's archaeological resources from the Paleo-Indian to the present: A survey of research problems and survey priorities*, ed. O. Chesler, 103–38. Office of Cultural and Environmental Services, New Jersey Department of Environmental Protection, Trenton.

Winfree, R. W.
- 1967 The T. Gray Hadden site, King William County, Virginia (44KW4). *Quarterly Bulletin of the Archaeological Society of Virginia* 22:2–26.
- 1969 Newington, King and Queen County. *Quarterly Bulletin of the Archaeological Society of Virginia* 23:160–224.

Winters, H. D.
- 1968 Value systems and trade cycles in the Late Archaic in the Midwest. In *New perspectives in archaeology*, ed. S. R. Binford and L. R. Binford, 175–221. Chicago: Aldine.

Witthoft, J.
- 1949 *Green corn ceremonialism in the Eastern Woodlands*. University of Michigan Museum of Anthropology, Occasional Contributions 13. Ann Arbor: University of Michigan Press.
- 1954 Pottery from the Stewart site, Clinton County, Pennsylvania. *Pennsylvania Archaeologist* 24(1):22–27.
- 1959 Ancestry of the Susquehannocks. In *Susquehannock Miscellany*, ed. J. Witthoft and W. F. Kinsey, 19–60. Harrisburg: Pennsylvania Historical and Museum Commission.

Witthoft, J., and S. S. Farver
- 1952 Two Shenks Ferry sites in Lebanon County, Pennsylvania. *Pennsylvania Archaeologist* 24(1):3–30.

Witthoft, J., W. F. Kinsey, and C. Holzinger
- 1959 A Susquehannock cemetery: The Ibaugh site. In *Susquehannock Miscellany*, ed. J. Witthoft and W. F. Kinsey, 99–119. Harrisburg: Pennsylvania Historical and Museum Commission.

Wittkovski, J. M.
- 1982a A summary of cultural resources and environmental variables on the Virginia Eastern Shore. *Quarterly Bulletin of the Archaeological Society of Virginia* 37:1–9.
- 1982b Eastern Shore archaeological survey summary. Virginia Research Center for Archaeology, Yorktown, Virginia. Typescript.

Wolfe, P.
- 1977 *The geology and landscapes of New Jersey*. Trenton, N.J.: Crane Russak.

Wolley, C.
1902 *A two years' journal in New York and parts of its territories in America*, ed. E. G. Bourne. Cleveland, Ohio: Burrows.

Wray, C. F.; and H. L. Schoff
1953 A preliminary report on the Seneca sequence in western New York, 1550–1687. *Pennsylvania Archaeologist* 23(2):53–63.

Wright, H. T.
1973 *An archaeological sequence in the middle Chesapeake region, Maryland.* Maryland Geological Survey. Archaeological Series no. 1. Baltimore.

Wright, L. B. and V. Freund, eds.
1953 *The history of travell into Virginia Britania, by William Strachey.* Hakluyt Society, Second Series no. 103. London: Hakluyt Society.

Zeisberger, D.
1910 History of the American Indians, edited by A. B. Hulbert and W. N. Schwauze. *Ohio State Archaeological and Historical Quarterly* 19:1–189.

Zukerman, K.
1979a Slaughter Creek comprehensive survey, phase I. Island Field Museum, South Bowers, Delaware. Typescript.
1979b Slaughter Creek comprehensive survey, phase II. Island Field Museum, South Bowers, Delaware. Typescript.

Notes on Contributors

MARSHALL J. BECKER is associate professor of anthropology at West Chester University. His research interests include the ethnohistory of the Lenape.

Jay F. Custer is associate professor of anthropology and director of the Center for Archaeological Research, Department of Anthropology, University of Delaware. He is interested in prehistoric archaeology of the Delmarva region and eastern North America.

DANIEL R. GRIFFITH, chief of the Delaware Bureau of Archaeology and Historic Preservation, studies prehistoric ceramics, lithics, settlement patterns, and prehistoric archaeology of eastern North America.

CHRIS C. HUMMER is a Ph.D. candidate at Temple University. His research interests include Woodland period archaeology in eastern North America, lithics, early ceramics, and socialization.

HERBERT C. KRAFT, a professor of anthropology, director of the Seton Hall University Museum, and director of the Archaeological Research Center at Seton Hall, studies and writes on northeastern and European archaeology.

R. MICHAEL STEWART is currently senior archaeologist in the Cultural Resource Group of Louis Berger and Associates, Inc., East Orange, New Jersey. Dr. Stewart researches lithics, settlement patterns, northeastern archaeology, soils, and paleoenvironments.

E. RANDOLPH TURNER, senior prehistoric archaeologist at the Virginia Research Center for Archaeology in Yorktown, is interested in prehistory and ethnohistory of eastern North America.

Index

Abbott Farm site (New Jersey), 107
Abbott Farm site complex (New Jersey), 67; cultigens, 78–79; Excavation 14, 69, 70, 72, 76, 78, 80; macro-band camp, 71–72; settlement models, 70–78; social organization, 79–80; stations, 73; subsistence, 78–79; transient camp, 72–73
Abbott's Brook site (New Jersey), 72
Abbott's Lane site (New Jersey), 93
Abbott zone-decorated ceramics, 69
Accohannoc Indians, 53, 54, 150, 164
Accomac Indians, 53, 54, 150, 164
Agricultural intensification, 160–65
Algonkian language, 34, 50, 52, 102, 103
Allard, John, 111
Amaranth, 38, 46, 65
Anadromous fish, 67, 80, 88, 106–7
Andaste Indians. See Susquehannock complex
Appalachian Highlands culture summary, 157–58
Appomattox River, 24
Archaeological Society of Maryland, 36, 39
Armewamex Indians, 105
Arrowhead Farm site (Maryland), 61
Assateague Indians, 52, 53, 54
Atlantic Coast, 35, 36, 38, 40, 41, 46, 50
Atsayonck Indians, 105

Bainbridge incised ceramics, 118, 123
Band level organizations: composite bands, 144; definitions, 144; patrilocal bands, 144
Barkers Landing complex, 57
Bay Vista site (Delaware), 30, 38, 49
Beaver Creek, 83
Bell Philower site (New Jersey), 103
Big House Church, 115
Big man organization, 80, 88
Blue Rock Phase. See Shenks Ferry complex
Blue Rock site (Pennsylvania), 123, 129, 130
Book Mound site (Pennsylvania), 117, 118
Bordentown Water Works site (New Jersey), 72
Bowmans Brook ceramics, 59, 70
Brandywine band. See Lenape Indians
Brandywine Creek, 99

Brinsfield Island site (Maryland), 36
Brock Mound site (Pennsylvania), 118
Brock Village site (Pennsylvania), 118
Buckingham site (Maryland), 35
Budd, Thomas, 102
Bull Run site (Pennsylvania), 118, 122, 131
Bushy head mask, 115
Byram site (Pennsylvania), 84

Canarsee Indians, 105
Cape Henlopen spit, 37, 41, 42, 53, 55, 151
Carey complex, 35, 37, 42
Carney Rose site (New Jersey), 72
Castle Creek ceramics, 86
Ceramics: ethnic groups correlations, 116; social interactions, 117
Chance incised ceramics, 125
Charnel house, 54
Chenopodium, 38, 46, 65
Chesapeake Bay, 36, 39, 40, 41, 42, 46, 51, 53, 55, 61, 119, 136, 152
Chester River, 41
Chicamacomico River, 36
Chickahominy River, 24
Chicone sites (Maryland), 39
Chiefdoms: archaeological characteristics, 20–21; definitions, 19–21, 145–47; origins, 167–68; types, 22–23, 146
Chincoteague Bay, 40
Choptank Indians, 52, 53, 151, 164
Choptank River, 34, 39, 50, 53
Christina River, 62, 63, 92
Circumscription, 151
Clemson Island ceramics, 117, 122
Clemson Island complex, 117–19; cultigens, 118; settlement pattern, 118; subsistence, 118
Clemson Island Mound site (Pennsylvania), 117, 118
Clyde Farm site (Delaware), 62, 63, 64
Cognatic tribes, 145, 153, 157, 164, 165
Composite bands, 144, 147
Composite tribes, 145
Conowingo site (Maryland), 119
Crane Hook site (Delaware), 61, 64
Crum Creek, 97

Index

Custaloga's Town, 94
Custer, Jay F., 27–28

Dankers, Jasper, 109
Dark Moon site (New Jersey), 83
Delaware Bay, 36, 38, 40, 46, 49, 50, 52, 53, 56
Delaware Bureau of Archaeology and Historic Preservation, 35
Delaware Indians, 52, 58, 73, 78, 86, 90, 106
Delaware Park site (Delaware), 64, 65
Delaware Water Gap, 102
Delmarva Adena complex, 51
Derrockson site (Delaware), 38
Descent line social systems, 166
Dutch settlers, 96, 99, 102

Eastern Neck Island, 39
Eastern Shore. *See* Maryland Eastern Shore; Virginia Eastern Shore
East River, 24
East River ceramics, 86, 87
Egypt Road site, 36
Elk River, 61
English settlers, 99, 102
Erb Rockshelter site, 119
Esopus Indians, 102, 105
Evolutionary theory, 165–68

Fairmont site (Maryland), 39
Fisher Farm site (Pennsylvania), 117, 118, 122, 123, 130
Fish weirs, 107
Fleming site (Maryland), 39
Forks of the Delaware, 93
Fort Ancient ceramics, 123
Fort Ancient complex, 159
Fowling Creek site (Maryland), 36
Fried, Morton, 28, 142
Fuert Focus, 123
Funk phase. *See* Shenks Ferry complex
Futer, Arthur, 138

Golden Island site (Maryland), 39
Graybill, Jeffrey, 11, 127–29
Great Neck site (Virginia), 25
Green Valley site complex (Delaware), 63, 64
Grey's Run Rockshelter site (Pennsylvania), 118, 122
Gropp's Lake site (New Jersey), 69, 72, 73

Hackensack Indians, 102, 105
Hagerstown Valley, 123

Harland, George, 99–100
Harry's Farm site (New Jersey), 108, 111
Hatch site (Virginia), 25
Haverstraw Indians, 105
Hell Island ceramics, 119
Hell Island site (Delaware), 30, 31, 61, 64
Herring Island site (Maryland), 61
Hessian Run site (New Jersey), 62
Holiday Park site (Maryland), 39
Hollingsworth Farm site (Maryland), 61, 64
Homogeneous tribes, 145
Household cluster, 130
Hughes-Willis site (Delaware), 35, 41, 46, 47
Hunters Home complex, 117
Huskanawing, 22

Independence Mall site (New Jersey), 73
Indianhead ceramics, 70
Indianhead site (New Jersey), 61, 64
Indian Knoll Landing site (Maryland), 39
Indian Landing site (Delaware), 38, 45, 46, 47
Indian River Bay, 38
Industrial Terrace site (New Jersey), 71, 72, 79
Iroquoian ceramics, 69
Iroquois Indians, 11, 134, 136, 140, 147, 153, 159–60
Iroquois language, 103
Island Field site (Delaware): oyster utilization, 98; Woodland I occupation, 31; Woodland II occupation, 37–38, 41, 47, 48, 49
Ithaca linear ceramics, 125

James River, 21, 24, 25
Jamestown, Virginia, 22
Jersey Indians, 92, 93, 100
Juniata River, 130

Kelso corded ceramics, 115, 118, 123
Kent, Barry C., 116, 134
Kent County Archaeological Society, 35
Keyhole structures, 130
Keyser Farm ceramics, 51, 123
Kibbler-Funk site (Pennsylvania), 132
King Cole site (New Jersey), 82
Kipp Island No. 4 site (New York), 118
Kitchtawank Indians, 105
Kittatinny Mountains, 106, 108

Laet, Johan de, 109
Lalor's Field site (New Jersey), 71, 78, 79
Lambertsville site (New Jersey), 82, 84, 85
Lancaster phase. *See* Shenks Ferry complex

Lankford site (Maryland), 36, 45
Late Carey complex, 30–31, 42
Lenape Indians, 141, 144, 148, 149, 151, 161; Brandywine band, 98–100; cultigens, 76; Okehocking band, 94–98; social organization, 93–95
Levanna ceramics, 118, 123
Lewes High School site (Delaware), 38
Lighthouse site (Delaware), 37
Lineal tribes, 145, 153, 165
Lister site (New Jersey), 72
Little Ice Age, 135–36
Locust Grove ceramics, 125
Long Point site (Maryland), 36
Lower Black Eddy site (Pennsylvania), 82
Lower Delaware Valley culture summary, 148–50
Lower Delmarva Peninsula culture summary, 150–53
Lower Geanquakin site (Maryland), 39
Luray Focus, 156

McFate ceramics, 125
Maloney site (Maryland), 39
Mantaes Indians, 105
Marshyhope Creek, 39
Maryland Eastern Shore, 29, 36, 40, 51
Maryland Geological Survey, Division of Archaeology, 36, 39
Mason Island complex, 156
Massachusett Indians, 27
Massapequa Indians, 105
Matinocock Indians, 105
Matrilineages, 147, 150
Mattaponi River, 23, 24, 26
Maximal chiefdom, 146
Mesingw, 115
Middle Delaware River culture summary, 148–50
Middle Delaware site complex: cultigens, 84–85; settlement patterns, 82–85; social complexity, 86–87; subsistence, 85–86
Miller Field site (New Jersey), 108, 111, 113, 115
Miller-Toms site (Delaware), 38
Millman site complex (Delaware), 37
Minguannan ceramics, 30, 51, 59
Minguannan complex, 119, 124; diagnostic artifacts, 59; ethnohistoric data, 65–67; macroband base camps, 61–62; microband base camps, 62–63; procurement site, 63; settlement patterns, 59–64; storage features, 64; subsistence, 65

Minguannan site (Pennsylvania), 62, 64, 115
Minimal chiefdom, 146
Minisink Indians, 85, 102, 104
Minisink phase, 104; burial data and ideology, 113–15; cultigens, 112; European trade goods, 114; longhouses, 109; settlement and community patterns, 108–13; storage pits, 109, 111–12; subsistence, 106–8, 112
Minisink site (New Jersey), 108
Minquas. *See* Susquehannock complex
Mispillion River, 41, 42, 52
Mispillion Site (Delaware), 30, 35, 41, 45, 50, 55
Mississippian cultures, 159
Mississippi River Valley, 160
Mitchell Farm site (Delaware), 62, 64
Mockley ceramics, 119
Monocacy River, 123, 156
Monocan Indians, 157
Monogahela ceramics, 123
Monongahela complex, 91, 159, 164
Montgomery Focus, 123, 156
Moore site (Maryland), 36
Moravian missionaries, 102
Moyaone ceramics, 59
Multilinear evolution, 12, 28, 143
Munsee incised ceramics, 125
Munsee Indians, 91, 93, 144, 150, 163; proto-Munsee, 105–6
Munsee language, 102, 105
Murderkill River, 40
Murry site (Pennsylvania), 132–33, 137, 142

Nace site (Pennsylvania), 118–19, 123, 129, 130, 133
Nanticoke Indians, 52, 53, 151, 164
Nanticoke River, 36, 39
Naraticonck Indians, 105
Narragansett Indians, 27
Nassawango site (Maryland), 36
Navasink Indians, 105
Newport site (Delaware), 61, 64
Nochpeem Indians, 105
Northbrook site (Pennsylvania), 98
North Carolina, 21, 158
North Museum, 128, 138
Nyack Indians, 109

Occohannock site (Virginia), 36
Ohio River, 160
Okehocking band. *See* Lenape Indians
Opechancanough, 53

Index

Optimum foraging theory, 48
Ossuaries: ethnohistoric sources, 54; Powhatan chiefdom, 25; Slaughter Creek complex, 34, 50; Slaughter Creek site, 34
Overpeck ceramics, 30, 51, 59, 70, 82, 87, 89
Overpeck site (Pennsylvania), 82, 84, 85, 86
Owasco ceramics, 70, 86, 87, 103–4, 117, 118, 119, 122, 123

Page ceramics, 123
Pahaquarra phase, 104
Pahaquarra site (New Jersey), 108, 113, 114
Pamunkey River, 23, 24, 25
Patawomeke site (Virginia), 24, 25
Patrilocal band, 144
Penn, William, 92, 96, 100
Pine Bluff site (Maryland), 39
Pit houses, 33, 35, 37–38, 65
Pocomoke Indians, 53
Pocono Mountains, 106, 108
Point Peninsula complex, 117
Polynesian societies, 166–67
Poplar Thicket site (Delaware), 30, 38, 49
Potomac Creek ceramics, 51
Potomac River, 11, 24, 25, 34, 52, 56
Potomac Valley culture summary, 154–56
Powhatan Chiefdom, 21–26, 53, 150, 152, 159, 164; internal organization, 22; size, 21–22; status ranking, 22–23
Prickly Pear Island site (Delaware), 39
Pulpwood Landing site (Maryland), 39

Quioccosan, 54

Racconn Point site (New Jersey), 61, 64
Rachel site (Maryland), 39
Ramified societies, 166
Ranked societies, 28
Rappahannock River, 25
Raritan Indians, 102, 105
Rechgawawanks Indians, 102, 105
Reeves site (Maryland), 39
Rehoboth Midden site (Delaware), 37
Remkokes Indians, 105
Ridley Creek, 97
Riggins ceramics, 69
Ritter site (Delaware), 38, 46
Riverview Cemetery site (New Jersey), 72, 78
Robbins Farm site (Delaware), 51
Rockaway Indians, 105
Roebling Park site (New Jersey), 71, 76, 78

Rosencrans site (New Jersey), 113
Russell site (Delaware), 35

Sahlins, Marshall D., 166–67
Saint Jones Neck, 37, 40
Saint Jones River, 52
Saint Lawrence River, 136
Saint Martin site (Maryland), 39
Sandt's Eddy site (New Jersey), 82
Sandy Point site (Maryland), 39
Sanhican Indians, 105
Scacht site (Pennsylvania), 135
Schultz incised ceramics, 70, 125
Schultz site (Pennsylvania), 136, 137, 140
Schuylkill Indians, 105
Schuylkill River, 92, 94, 97
Segmented tribes, 145
Service, Elman, 28, 144
Sewapois Indians, 105
Shady Brook site (New Jersey), 69, 70, 72, 76, 78
Sheep Rock Shelter site (Pennsylvania), 118, 130
Shellfish utilization, 47–49
Shenks Ferry ceramics, 117, 121–25
Shenks Ferry complex, 11, 12, 62, 116, 153, 156, 163, 164; Blue Rock phase, 119, 124, 125, 126, 128, 129, 131, 163; burials, 133; chronology, 121; community patterns, 129–34; cultigens, 133; Funk phase, 124, 125, 126, 129, 133, 134, 135, 142, 163; Lancaster phase, 124, 125, 126, 129, 132, 133, 134, 135, 142, 163; origins, 118–20, 124; procurement sites, 129–30; relationship with Susquehannock complex, 134–38; sedentary villages, 132–34; settlement patterns, 125–29; social organization, 133–34; Stewart phase, 131; subsistence, 131, 133
Shepard ceramics, 123
Siconese Indians, 105
Sinepuxent Neck sites (Maryland), 49
Slaughter Creek site (Delaware), 30, 33–34, 41, 45, 50
Sluyter, Peter, 109
Smith, John, 23
Smith Island middens (Maryland), 39
Spring Branch site (Maryland), 36
Stasl site (Maryland), 39
Steelman site (Maryland), 39
Stewart, R. Michael, 28, 123
Strickler site (Pennsylvania), 138
Sturgeon Pond site (New Jersey), 73

Susquehanna Valley culture summary, 153–54
Susquehannock complex, 11, 62, 70, 92, 116, 124, 153; cultigens, 138, 142; ethnohistoric data, 134; grave goods, 141–42; population, 137; relationship with Shenks Ferry complex, 134–38; settlement pattern, 137–38; social organization, 140–42; subsistence, 138–40
Sussex Society for Archaeology and History, 34
Swartswood Lake site (New Jersey), 108
Swedish settlers, 96, 98, 99

Tappan Indians, 105
Tennessee, 158
Teshoa, 112
Thompson's Island site (Delaware), 34, 50
Thousand Acre Rockshelter site (New Jersey), 82
Tizzard Island site (Maryland), 36
Townsend ceramics, 34, 35, 51, 59; definition, 29–30
Townsend site (Delaware), 34–35, 41, 45, 49, 50
Triangular projectile points, 59, 72, 78, 82–83
Tribal organizations: definitions, 144–45; development, 165–67; varieties, 145
Tribelet, 165
Tuckahoe site (Maryland), 36
Tuckahoe Spring site (Maryland), 36
Turkey Swamp site (New Jersey), 79
Typical chiefdom, 146

Unalachtigo, 93
Unami Creek Rockshelter site (Pennsylvania), 82
Unami language, 53, 58, 93, 105, 106
University of Delaware, Department of Anthropology, 38
Upland Victorian site (Delaware), 62
Upper Bare Island Rockshelter site (Pennsylvania), 132
Upper Black Eddy site (Pennsylvania), 82
Upper Delaware Valley culture summary, 146–48
Upper Delmarva Peninsula culture summary, 148–50

Vinette I ceramics, 117
Virginia Coastal Plain culture summary, 156–57
Virginia Eastern Shore, 29, 39, 40
Virginia Research Center for Archaeology, 36
Visscher, Nicholas, 111

Walking Purchase Confirmation Treaty, 100
Wallpack Bend, 104
Walter's Nursery site (New Jersey), 82
Wampanoag Indians, 27
Wappinger Indians, 105
Warranawankong Indians, 105
Warrington site (Delaware), 30, 38, 49
Washington Boro Village site (Pennsylvania), 138
Washington's Crossing site (Pennsylvania), 82, 85
Watson House site, 71, 72
Webb complex, 30–31, 42, 51, 151, 161
Webb site (Pennsylvania), 62, 64
Wells site (New York), 117
Wessell site (Maryland), 36
West Branch Valley, Susquehanna River, 117, 118, 122
White, John, 111
White Clay Creek, 61, 62
White Horse site (New Jersey), 73
Wiechquaeeskeck Indians, 105
Wilgus site (Delaware), 37, 45, 46, 47, 48, 49
William Penn Memorial Museum, 128
Williamson site (New Jersey), 79, 82, 83–84, 85, 86, 87
Willin site (Maryland), 39
Witthoft, John, 138
Wolf Run Fort site (Pennsylvania), 131
Woodbury Annex site (New Jersey), 61, 64
Woodland cord-marked ceramics, 119
Woods site (Delaware), 63
Wright's Field site (New Jersey), 78, 79
Wynicaco, 54
Wyoming Valley ceramics, 125, 137
Wyoming Valley complex, 135

York River, 21, 24

974
LAT

R15

Late woodland cultures
of the Middle
Atlantic region